Lecture Notes in Mathematics

Edited by A. Dold and B. Eckmann

1074

Edward W. Stredulinsky

Weighted Inequalities and Degenerate Elliptic Partial Differential Equations

Springer-Verlag
Berlin Heidelberg New York Tokyo 1984

Author

Edward W. Stredulinsky
Department of Mathematics, University of Wisconsin
Madison, Wisconsin 53706, USA

AMS Subject Classification (1980): 35J70

ISBN 3-540-13370-4 Springer-Verlag Berlin Heidelberg New York Tokyo
ISBN 0-387-13370-4 Springer-Verlag New York Heidelberg Berlin Tokyo

Printing and binding: Beltz Offsetdruck, Hemsbach/Bergstr.
2146/3140-543210

TABLE OF CONTENTS

INTRODUCTION

The main purpose of these notes is the analysis of various weighted spaces and weighted inequalities which are relevant to the study of degenerate partial differential equations. The usefulness of these results is demonstrated in the later part of the text where they are used to establish continuity for weak solutions of degenerate elliptic equations.

The most important inequalities dealt with are certain weighted Sobolov and Poincaré inequalities for which the admissible weights are characterized. Weighted reverse Hölder inequalities and weighted inequalities for the mean oscillation of a function are dealt with as well. A much larger class of degeneracies is considered than previously appears in the literature and some of the applications are known only in the strongly elliptic case.

Two approaches are taken to the problem of establishing continuity of weak solutions. The first approach taken involves a Harnack inequality and the second Morrey spaces. The first applies to equations of the form $\operatorname{div} A = B$, where A, B satisfy the growth conditions

$$|A| \leq \mu(x)|\nabla u|^{p-1} + a_1(x)u^{p-1} + a_2(x) ,$$

$$A\cdot\nabla u \geq \lambda(x)|\nabla u|^p - c_1(x)u^p - c_2(x) ,$$

$$|B| \leq b_0\lambda(x)|\nabla u|^p + b_1(x)|\nabla u|^{p-1} + b_2(x)u^{p-1} + b_3(x) .$$

A Harnack inequality is proved for weights μ, λ satisfying certain capacitary conditions. Interior continuity follows immediately from this, and a Wiener criterion is established for continuity at the boundary. This generalizes work of D. E. Edmunds and L. A. Peletier [EP], R. Gariepy and W. P. Ziemer [GZ], S. N. Kruzkov [K], M. K. V. Murthy and G. Stampacchia [MS], P. D. Smith [SM], and N. S. Trudinger [T1].

A theory of weighted Morrey spaces is developed which establishes continuity estimates for a wide class of weighted Sobolev spaces $W^{1,p}$ with $p > d$, d the spatial dimension. This is in turn applied to prove the continuity of solutions of systems of the form $\operatorname{div} A_i = B_i$, $i = 1,\dots,m$, where A_i and B_i satisfy growth conditions similar to the above with $p > d - \varepsilon$. In the non-degenerate case this is due to K. O. Widman [WI] and, in more general form to N. G. Meyers and A. Elcrat [MYE].

It is necessary to mention related work of E. B. Fabes, C. E. Kenig, D. S. Jerison, and R. P. Serapioni [FKS], [FJK] which was done independently at the same time as the work presented in these notes. The approach taken and the material covered differ considerably but there is a certain overlap (see comments before 2.2.40 and the introduction to Section 3.1.0.).

The following is a brief description of the contents of each chapter. The reader interested mainly in the applications should proceed immediately to Chapter 3.

Chapter 1 contains the basic analysis needed for Chapter 2. The relationship between maximal functions, covering lemmas and Lebesgue differentiation of integrals is reviewed. A calculus for functions absolutely continuous with respect to a measure is developed and the admissible weights for several new variations of Hardy's inequalities are characterized. Finally, several comparability results are proved for "capacities" and set functions which appear later in the analysis of the weighted Sobolev inequalities.

Chapter 2 deals mainly with weighted Sobolev inequalities and properties of weighted Sovolev spaces. The characterization of weights for Sobolev inequalities is carried out in a very general setting in the first section and is translated to a more useful form in Section 2.2.0 where, in addition, some examples are developed. The main thrust of Section 2.2.0 however, is the development of results relating capacity, quasicontinuity, convergence in weighted Sobolev spaces and weak boundary values for Sobolev functions. Chapter 2 closes with a result on weighted reverse Hölder Inequalities.

All direct applications to differential equations are contained in Chapter 3. These include the Harnack inequality as well as the interior and boundary continuity results for weak solutions of divergence type degenerate elliptic equations (3.1.0); modulus of continuity estimates for Sobolev functions and functions of vanishing mean oscillation (3.2.0); and the continuity result for weak solutions of degenerate elliptic systems in a "borderline" case (3.3.0).

I would like to express my sincere thanks to William Ziemer under whose guidance this work was completed. I would also like to thank David Adams, John Brothers, and Alberto Torchinsky for conversations pertaining to this material.

CHAPTER 0

The following is a short list of conventions and notation to be used throughout the text.

Sections, theorems, and statements each are labelled with a sequence of three numbers, the first two denoting the chapter and section, the third denoting order within the particular section.

The Lebesgue measure of a set E is represented as $|E|$. H^n represents n-dimensional Hausdorff measure. The abbreviations sup, inf will be used to represent the essential supremum and infimum unless it is specified otherwise. $B(x,r)$ is the open ball of radius $r > 0$, centered at x. The specific space in which $B(x,r)$ is contained will be clear from the context. Sometimes the notation $B_r = B(x,r)$ is used. χ_E is the characteristic function of the set E, that is, $\chi_E(x) = \begin{cases} 1 & x \in E \\ 0 & \text{otherwise} \end{cases}$. The letter c will be used to represent constants which may differ from line to line but which remain independent of any quantities of particular importance to the specific calculation being carried out. $L^p(\omega,E)$ is the space of equivalence classes of measurable functions $u : E \to R$ such that $\int |u|^p d\omega < \infty$. Finally, ∇u denotes the gradient of u, that is $\nabla u = (\frac{\partial u}{\partial x_1}, \ldots, \frac{\partial u}{\partial x_d})$.

CHAPTER 1

The results of Chapter 1 are of little direct interest from the point of view of differential equations but are necessary tools in proofs of the major theorems of Chapters 2 and 3. 1.1.1 and 1.1.8 deal with the relationship between covering properties, maximal functions, and the differentiation of integrals. The basic calculus for functions absolutely continuous with respect to a measure is developed in 1.1.10. In Section 1.2.0 the weights for several variations of Hardy's inequalities are characterized, and in Section 1.3.0 a number of capacities and set functions are shown to be comparable.

1.1.0 Calculus in Measure Spaces

The basic motivation for the inclusion of Section 1.1.0 is the desire to present in an elementary manner special cases of known results which are needed in other sections.

1.1.1 Covering Properties, Maximal Functions, and Differentiation of Integrals. Let Ω be a topological space and (Ω,ω) a measure space with ω positive such that the integrable continuous functions are dense in $L^1(\omega,\Omega)$. For instance, this is true if ω is a locally finite regular Borel measure and Ω is a σ-compact Hausdorff space. Recall also that every locally finite Borel measure on $\mathbf{R}^n$ is regular.

Let $\mathbf{H} = \bigcup_{y\in\Omega} \mathbf{H}_y$, where $\mathbf{H}_y$ is a nonempty collection of measurable sets B with $y \in B$ and $0 < \omega(B) < \infty$ and

$$Mf(y) = \sup_{B\in \mathbf{H}_y} \frac{1}{\omega(B)} \int_B |f| d\omega .$$

It is said that M satisfies a weak L^1 estimate if there exists $c_1 > 0$ such that

$$\omega(\{Mf > \lambda\}) < \frac{c_1}{\lambda} \int |f| \, d\omega$$

for all $f \in L^1(\omega,\Omega)$.

Consider the following covering property for some collection $\{\mathbf{L}_y\}_{y\in\Omega}$:

1.1.2 There exists $c_1 > 0$ such that if $E \subseteq \Omega$ is measurable and $\mathbf{G} \subseteq \bigcup_{y\in\Omega} \mathbf{L}_y$ is a cover of E such that for every $y \in E$ there exists $B \in \mathbf{G} \cap \mathbf{L}_y$, then there exists $\mathbf{F}$, an at most countable collection of pairwise disjoint sets, such that $\mathbf{F} \subseteq \mathbf{G}$ and $\omega(E) < c_1\omega(\bigcup_{\mathbf{F}} B)$.

1.1.3 Proposition. If $\{\mathbf{H}_y\}_{y\in\Omega}$ satisfies property 1.1.2, then M satisfies a weak L^1 estimate.

1.1.4 Proposition. If $\omega(\Omega) < \infty$, $1 < p < \infty$ and M satisfies a weak L^1 estimate, then

$$\int (Mf)^p d\omega < c_2 \int |f|^p d\omega$$

for all $f \in L^p(\omega,\Omega)$, where $c_2 = \frac{2p}{p-1} c_1$.

1.1.5 Proposition. If M satisfies a weak L^1 estimate and $f \in L^1(\omega,\Omega)$, then

$$\lim_{\alpha\to 0} \sup\Big\{\frac{1}{\omega(B)} \int_B |f - f(y)|d\omega : B \in H_y \,, \text{ diam } B < \alpha\Big\} = 0 \qquad (1.1.6)$$

for almost all $y \in \Omega$. The convention is used that the supremum taken over the empty set is zero.

1.1.7 Remark. (1.1.6) implies that $f(y) \le Mf(y)$ almost everywhere if H_y contains sets of arbitrarily small diameter.

Proof of 1.1.3. Assume $\{H_y\}_{y\in\Omega}$ satisfies 1.1.2 and let $E_\lambda = \{Mf > \lambda\}$. For each $y \in E_\lambda$, $\exists\ B \in H_y$ such that $\frac{1}{\omega(B)} \int_B |f|d\omega > \lambda$. Let G Be a covering of E_λ consisting of such sets and use 1.1.2 to get $\mathbf{F} \subseteq \mathbf{G}$, $\mathbf{F}$, an at most countable collection of pairwise disjoint sets with $\omega(E_\lambda) \le c_1\omega(\bigcup_{\mathbf{F}} B)$, so that

$$\omega(E_\lambda) \le c_1 \sum_{\mathbf{F}} \omega(B) \le \frac{c_1}{\lambda} \sum_{\mathbf{F}} \int_B |f|d\omega \le \frac{c_1}{\lambda} \int_\Omega |f|d\omega \ . \ \blacksquare$$

Proof of 1.1.4. Given $f \in L^p(\omega,\Omega)$, $1 < p < \infty$, it follows that $f \in L^1(\omega,\Omega)$ since $\omega(\Omega) < \infty$. Without loss of generality, assume $f \ge 0$. Let $f_\lambda = \chi_{\{f>\lambda/2\}}f$ so that $f \le f_\lambda + \lambda/2$ and $Mf \le Mf_\lambda + \lambda/2$, but then $\omega(\{Mf > \lambda\}) \le \omega(\{Mf_\lambda > \lambda/2\})$ $\le \frac{2c_1}{\lambda} \int f_\lambda d\omega = \frac{2c_1}{\lambda} \int_{\{f>\lambda/2\}} f\, d\omega$ and

$$\int Mf^p d\omega = p \int_0^\infty \lambda^{p-1}\omega(\{Mf > \lambda\})d\lambda \le 2pc_1 \int_0^\infty \lambda^{p-2} \int_{\{f>\lambda/2\}} f\, d\omega\, d\lambda$$

$$= 2pc_1 \int_\Omega f \int_0^{2f} \lambda^{p-2} d\lambda\, d\omega = \frac{2^p p}{p-1}\, c_1 \int_\Omega f^p d\omega \ . \ \blacksquare$$

Proof of 1.1.5. Let $Lf(y) = \lim_{\alpha\to 0} \sup\Big\{\frac{1}{\omega(B)} \int_B |f - f(y)|d\omega : B \in H_y,$ $\text{diam } B < \alpha\Big\}$ so $Lf(y) \le Mf(y) + |f(y)|$ and

$$\omega(\{Lf > \lambda\}) \le \omega(\{Mf > \lambda/2\}) + \omega(\{|f| > \lambda/2\}) \le \frac{2(c_1 + 1)}{\lambda} \int |f|d\omega \ .$$

If g is continuous and integrable, then it is clear that $Lg = 0$. Choose g_n continuous such that $g_n \to f$ in $L^1(\omega,\Omega)$. $Lf \le L(f - g_n)$ $+ Lg_n = L(f - g_n)$ and so $\omega(\{Lf > \lambda\}) \le \omega(\{L(f - g_n) > \lambda\})$ $\le \frac{2(c_1 + 1)}{\lambda} \int |f - g_n|d\omega \to 0$ as $n \to \infty$. Thus $Lf = 0$ almost everywhere. $\blacksquare$

1.1.8 Covering Lemmas

The covering lemma 1.1.9 is a direct generalization to doubling measures of a standard covering lemma for Lebesgue measure. For nondoubling measures this may be replaced by Besicovitch-type covering lemmas, a very general form of which is proved in [MR], the proof following the basic outline in Besicovitch's original paper [B]. A more accessible reference is [G].

1.1.9 Proposition. If ω is a doubling measure in a bounded open set Ω, i.e. $\omega(B(x,2r)) < c_\omega \omega(B(x,r))$ for all x, r such that $\bar{B}(x,2r) \subseteq \Omega$, then the covering property 1.1.2 holds with $\{L_y\}_{y\in\Omega}$ being the collection of all balls $B \subseteq \Omega$ with $y \in B$.

Proof. Proceed as in [ST], page 9.

1.1.10 Calculus for Functions Absolutely Continuous to a Measure. The basic calculus for functions absolutely continuous with respect to a measure ω closely resembles that for ω = Lebesgue measure.

If ω is a finite positive Borel measure on $[a,b)$ and $f\colon [a,b) \to \mathbf{R}$, then it is said that f is absolutely continuous with respect to ω if

$$\begin{cases} \forall\, \varepsilon > 0 \;\exists\, \delta > 0 \text{ so that if } \sum_{i=1}^{\infty} \omega(I_i) < \delta, \text{ where} \\ \text{the } I_i = [a_i,b_i) \subseteq [a,b) \text{ are pairwise disjoint} \\ \text{intervals, then } \sum_{i=1}^{\infty} |f(b_i) - f(a_i)| < \varepsilon . \end{cases} \tag{1.1.11}$$

As a direct consequence f is left-continuous and in fact discontinuous only on atoms of ω.

Let $N = \{y \in [a,b) : \omega[y,x) = 0 \text{ for some } x > y,\ x \in [a,b)\}$. N is a countable union of disjoint maximal intervals of measure zero, and so

$$\omega(N) = 0 . \tag{1.1.12}$$

The results of the previous section will be applied to the measure space $[a,b) - N$, with H_y consisting of all intervals $[y,x)$ with $x > y$, $x \in [a,b)$, so that $Mf(y) = \sup_{y<x<b} \frac{1}{\omega([y,x))} \int_{[y,x)} f\, d\omega$. The fact that continuous integrable functions are dense in $L^1(\omega,[a,b) - N)$ follows from this being true in $L^1(\omega,[a,b))$.

1.1.13 Proposition. Suppose f, g are absolutely continuous with respect to ω. Then:

$$f \text{ is of bounded variation and } f(t) = f(a) + P_a^t - N_a^t, \tag{1.1.14}$$

where P_a^t, N_a^t are the positive and negative variations of f on $[a,t)$.

(1.1.15) P_a^t, N_a^t are absolutely continuous with respect to ω and induce measures p, n absolutely continuous to ω so that

$$f(t) = f(a) + \int_{[a,t)} \frac{df}{d\omega}\, d\omega ,$$

where $\frac{df}{d\omega}$ is defined as $\frac{df}{d\omega} = \bar{p} - \bar{n}$ for $\bar{p}$, $\bar{n}$, the densities of p, n with respect to ω.

$$\lim_{x \to y^+} \frac{1}{\omega([y,x))} \int_{[y,x)} h\, d\omega = h(y) , \tag{1.1.16}$$

ω almost everywhere for $h \in L^1(\omega,[a,b))$, and so

$$\lim_{x \to y^+} \frac{f(x) - f(y)}{\omega([y,x))} = \frac{df}{d\omega}(y) , \tag{1.1.17}$$

ω almost everywhere.

(1.1.18) fg is absolutely continuous with respect to ω and

$$\frac{d(fg)}{d\omega} = \frac{df}{d\omega} g + \frac{dg}{d\omega} f_+ ,$$

ω almost everywhere, with f_+ representing the limit from the right of f. The asymmetry disappears if it is realized that $f_+(y) \neq f(y)$ only if $\omega(\{y\}) \neq 0$, in which case $\frac{df}{d\omega}(y) = \frac{f_+(y) - f(y)}{\omega(\{y\})}$, and so

$$\frac{d(fg)}{d\omega}(y) = \frac{(f_+(y)-f(y))g(y) + (g_+(y)-g(y))f(y) + (g_+(y)-g(y))(f_+(y)-f(y))}{\omega(\{y\})} .$$

(1.1.19) If $g > c > 0$ for some c, then $1/g$ is absolutely continuous with respect to ω and

$$\frac{d}{d\omega}\left(\frac{f}{g}\right) = \frac{\frac{df}{d\omega} g - \frac{dg}{d\omega} f}{g g_+} ,$$

ω almost everywhere.

$$\int_{[a,b)} \frac{dg}{d\omega} f_+ d\omega = fg\Big|_a^b - \int_{[a,b)} \frac{df}{d\omega} g\, d\omega . \tag{1.1.20}$$

(1.1.21) If $F : \mathbf{R} \to \mathbf{R}$ is differentiable, then $F \circ f$ is absolutely continuous and

$$\frac{d}{d\omega}(F \circ f) = \begin{cases} F' \circ f \dfrac{df}{d\omega}, & \text{a.e. where } f \text{ is continuous} \\ \dfrac{F \circ f_+ - F \circ f}{f_+ - f}, & \text{everywhere } f \text{ is discontinuous .} \end{cases}$$

A typical application of Proposition 1.1.13 is the evaluation of $\int_{[a,\infty)} (\omega([t,\infty)) + \lambda)^\alpha d\omega(t)$ for $\alpha > -1$ and $\lambda > 0$, where ω is a finite positive Borel measure on $[a,\infty)$ and $\omega((a,\infty)) \neq 0$.

Let $f(t) = \omega([t,\infty)) + \lambda$ so that f is absolutely continuous with respect to ω on any finite interval and $\frac{df}{d\omega} = -1$ by (1.1.17). Let $b' = \inf\{t \in [a,\infty) : \omega([t,\infty)) = 0\}$, b' possibly ∞, and choose b such that $a < b < b'$ and $\omega[t,\infty) > c > 0$ on $[a,b)$ for some c. Altering the function $F(t) = t^{\alpha+1}$ on $(-\infty, c/2)$ if necessary to insure that it is differentiable on $\mathbf{R}$, apply 1.1.21 to see that $F \circ f$ is absolutely continuous on $[a,b)$ with respect to ω and that

$$\frac{dF \circ f}{d\omega}(t) = \begin{cases} -(\alpha+1)(\omega([t,\infty)) + \lambda)^\alpha, & \text{if } t \text{ is not an atom of } \omega \\ \dfrac{(\omega((t,\infty)) + \lambda)^{\alpha+1} - (\omega([t,\infty)) + \lambda)^{\alpha+1}}{\omega(\{t\})}, & \text{if } t \text{ is an atom of } \omega . \end{cases}$$

By 1.1.15 it follows that

$$\begin{aligned} (\omega([b,\infty)) + \lambda)^{\alpha+1} &= (\omega([a,\infty)) + \lambda)^{\alpha+1} \\ &\quad - (\alpha+1) \int_{[a,b)-T_b} (\omega([t,\infty)) + \lambda)^\alpha d\omega(t) \\ &\quad + \sum_{t_i \in T_b} (\omega((t_i,\infty)) + \lambda)^{\alpha+1} - (\omega([t_i,\infty)) + \lambda)^{\alpha+1} , \end{aligned}$$

where T_b is the set of atoms of ω in $[a,b)$ and so

$$\begin{aligned} &\int_{[a,b)} (\omega([t,\infty)) + \lambda)^\alpha d\omega(t) \\ &\quad = \frac{(\omega([a,\infty)) + \lambda)^{\alpha+1} - (\omega([b,\infty)) + \lambda)^{\alpha+1}}{\alpha+1} \\ &\quad + \frac{1}{\alpha+1} \sum_{t_i \in T_b} [(\omega((t_i,\infty)) + \lambda)^{\alpha+1} - (\omega([t_i,\infty)) + \lambda)^{\alpha+1} \\ &\quad + (\alpha+1)(\omega([t_i,\infty)) + \lambda)^\alpha \omega(\{t_i\})] . \end{aligned}$$

Let $b \to b'$, recall that $\omega(b',\infty) = 0$, and make a few adjustments if b' is an atom to get

$$\int_{[a,\infty)} (\omega([t,\infty)) + \lambda)^{\alpha} d\omega(t) = \frac{(\omega([a,\infty)) + \lambda)^{\alpha+1}}{\alpha + 1}$$

$$+ \frac{1}{\alpha + 1} \sum_{t_i \in T} [(\omega((t_i,\infty)) + \lambda)^{\alpha+1} - (\omega([t_i,\infty)) + \lambda)^{\alpha+1}$$

$$+ (\alpha + 1)(\omega([t_i,\infty)) + \lambda)^{\alpha} \omega(\{t_i\})] ,$$

where T is the set of atoms of ω in $[a,\infty)$. A simple calculation shows that

$$- \text{sign } \alpha \int_{[a,\infty)} (\omega([t,\infty)) + \lambda)^{\alpha} d\omega(t) < - \frac{\text{sign } \alpha}{\alpha + 1} (\omega([a,\infty)) + \lambda)^{\alpha+1} .$$

<u>Proof of 1.1.14</u>. This is a slight variation on the standard proof to avoid the discontinuities of f. Let $\delta > 0$ be related to $\varepsilon = 1$, as in (1.1.11), so that $\sum_{i=1}^{\infty} |f(b_i) - f(a_i)| < 1$ if $\sum_{i=1}^{\infty} \omega(I_i) < \delta$. Since $\omega([a,b))$ is finite, ω has at most a finite number of atoms of measure larger than $\delta/2$. Let these be located at $x_1,\ldots,x_n$, $x_i < x_{i+1}$. Pick ε_i such that $\omega(x_i,x_i + \varepsilon_i) < \delta$, $i = 1,\ldots,n$, so given $x_i < y < x_i + \varepsilon_i$, then $\omega[y,x_i + \varepsilon_i) < \delta$, and so $|f(x_i + \varepsilon_i) - f(y)| < 1$. Pick a partition $a = a_0 < a_1 < \ldots < a_{m_1} = b$, which includes $\{x_i\}$ and $\{x_i + \varepsilon_i\}$ and for which $\omega((a_j,a_{j+1})) < \delta$, $j = 0,\ldots,m_1 - 1$.

Given a partition $a = b_0 < \ldots < b_{m_2} = b$, let $a = c_0 < \ldots < c_{m_3} = b$ be a refinement including both $\{a_i\}$, $\{b_i\}$ so that

$$\sum_{i=0}^{m_2-1} |f(b_{i+1}) - f(b_i)| < \sum_{i=0}^{m_3-1} |f(c_{i+1}) - f(c_i)|$$

$$= \sum_{j=1}^{n} |f(c_{x_j}) - f(x_j)| + \sum_{\substack{c_i \neq x_j \\ j=1,\ldots,n}} |f(c_{i+1}) - f(c_i)| ,$$

where c_{x_j} is the division pt to the immediate right of x_j and so

$$\sum_{i=0}^{m_2-1} |f(b_{i+1}) - f(b_i)| < \sum_{j=1}^{n} |f(x_j + \varepsilon_j) - f(x_j)|$$

$$+ \sum_{j=1}^{n} |f(x_j + \varepsilon_j) - f(c_{x_j})|$$

$$+ \sum_i \sum_{\substack{a_i < c_k < a_{i+1} \\ c_k \neq x_j}} |f(c_{k+1}) - f(c_k)|$$

$$\leqslant \sum_{j=1}^{n} |f(x_j + \varepsilon_j) - f(x_j)| + n + m_1 ,$$

which is independent of the partition $\{b_i\}$, so f is of bounded variation. ∎

Proof of 1.1.15. From the above, it follows that $f(t) =$ $f(a) + P_a^t - N_a^t$, where P_a^t, N_a^t are the positive and negative variation of f on $\lceil a,t)$. P_a^t will be shown to be absolutely continuous with respect to ω, and the same will follow for N_a^t by considering $-f$.

Given $\varepsilon > 0$, let δ be as in (1.1.11). If $\sum_{i=1}^{\infty} \omega(I_i) < \delta$ for $I_i = [a_i, b_i)$, pairwise disjoint intervals, pick a partition $a_i = c_{i,0} < \dots < c_{i,n_i} = b_i$ such that $P_{a_i}^{b_i} < \sum_{j=0}^{n_i - 1} (f(c_{i,j+1}) - f(c_{i,j}))^+ + \frac{\varepsilon}{2^i}$, where $x^+ = \begin{cases} x, & x > 0 \\ 0, & x \leqslant 0 \end{cases}$. It is now clear that P_a^t is absolutely continuous since

$$\sum_{i=1}^{\infty} |P_a^{b_i} - P_a^{a_i}| = \sum_{i=1}^{\infty} P_{a_i}^{b_i}$$

$$< \sum_{i=1}^{\infty} \sum_{j=0}^{n_i - 1} |f(c_{i,j+1}) - f(c_{i,j})| + \varepsilon$$

$$\leqslant 2\varepsilon ,$$

since

$$\sum_{i=1}^{\infty} \sum_{j=0}^{n_i - 1} \omega \lceil c_{i,j}, c_{i,j+1}) = \sum_{i=1}^{\infty} \omega(I_i) < \delta .$$

1.1.23. Since P_a^t, N_a^t are monotone increasing and left-continuous, they induce measures p, n. To see that these are absolutely continuous to ω, let E be a set such that $\omega(E) = 0$. Given $\varepsilon > 0$, pick $\delta > 0$, as in (1.1.27) (with f replaced by P_a^t) and V relatively open in $[a,b)$ such that $E \subseteq V$ and $\omega(V) < \delta$. $V = \bigcup_{i=1}^{\infty} I_i$, where the I_i are pairwise disjoint intervals, $I_i = (a_i, b_i)$, $i = 2, \dots, \infty$ and either $I_1 = \lceil a_1, b_1)$ and $a_1 = a$ or $I_1 = (a_1, b_1)$. Only the first case will be dealt with, the other being similar.

$$p(V) = P_a^{b_1} + \sum_{i=2}^{\infty} P_a^{b_i} - (P_a^{a_i})_+$$

$$< P_a^{b_1} + \left(\sum_{i=2}^{\infty} P_a^{b_i} - P_a^{a_i+\delta_i}\right) + \varepsilon \quad \text{for some} \quad \delta_i > 0$$

$$< 2\varepsilon ,$$

since $\omega([a,b_1)) + \sum_{i=2}^{\infty} \omega[a_i + \delta_i, b_i) < \omega(V) < \delta$. The absolute continuity of n follows similarly.

The Radon-Nikodym theorem now gives the existence of $\bar{p}$, $\bar{n}$, the densities of p, n with respect to ω. Letting $\frac{df}{d\omega} = \bar{p} - \bar{n}$, 1.1.15 follows from (1.1.14) and (1.1.23). ∎

<u>Proof of (1.1.16).</u> Recalling (1.1.12), it is only necessary to show that $\{H_y\}_{y\in[a,b)-N}$ satisfies property 1.1.2 in order to use Proposition 1.1.5 to conclude for $h \in L^1(\omega,[a,b))$ that

$$\lim_{x\to y^+} \frac{1}{\omega[y,x)} \int_{[y,x)} h\, d\omega = h(y) \quad \text{a.e.} \quad \omega .$$

This being proven, (1.1.17) follows easily since

$$\frac{f(x) - f(y)}{\omega[y,x)} = \frac{1}{\omega[y,x)} \int_{[y,x)} \frac{df}{d\omega}\, d\omega$$

from 1.1.15.

To show property 1.1.2 is satisfied, let $\mathbf{G}$ be a collection of intervals $[y_\alpha, x_\alpha)$, $\alpha \in A$, A some index set.

<u>1.1.24.</u> It will be said that $\mathbf{F}$ is subordinate to $\mathbf{G}$ if $\mathbf{F} \subseteq \mathbf{G}$ <u>and</u> $\bigcup_{I\in\mathbf{F}} I = \bigcup_{I\in\mathbf{G}} I$.

<u>1.1.25.</u> It will be shown that there is an at most countable collection $\mathbf{F}$ subordinate to $\mathbf{G}$, in which case there is a finite collection $\mathbf{F}_1 \subseteq \mathbf{F}$ such that

$$\omega\Big(\bigcup_{I\in\mathbf{F}} I\Big) < 2\omega\Big(\bigcup_{I\in\mathbf{G}} I\Big) . \qquad (1.1.26)$$

Due to the properties of intervals, there then exists a collection $\mathbf{F}_2$ subordinate to $\mathbf{F}_1$ which has the property that every point in

$$\bigcup_{I\in\mathbf{F}_1} I = \bigcup_{I\in\mathbf{F}_2} I \qquad (1.1.27)$$

is covered at most twice by intervals in $\mathbf{F}_2$. $\mathbf{F}_2$ can then be split into two collections $\mathbf{F}_3$, $\mathbf{F}_4$, with $\mathbf{F}_2 = \mathbf{F}_3 \cup \mathbf{F}_4$, where the intervals in F_i, $i = 3,4$, are pairwise disjoint. Then

$$\omega\Big(\bigcup_{I\in F_2} I\Big) \le 2\omega\Big(\bigcup_{I\in F_i} I\Big) \qquad (1.1.28)$$

for one of $i = 3,4$, in which case considering 1.1.24 to (1.1.28), it is seen that property 1.1.2 is satisfied with $c_1 = 4$.

To show that there exists an at most countable collection F subordinate to G, let Y be the set of points y_α, $\alpha \in A$, such that y_α is not in the interior of any interval $[y_\beta, x_\beta)$, $\beta \in A$. It is claimed that Y is at most countable. For $y_\alpha \in Y$, no point of $[y_\alpha, x_\alpha)$ can lie in Y. Pick a rational number $r_\alpha \in [y_\alpha, x_\alpha)$ and pair it with y_α. r_α cannot be paired in this way with any other element of Y, so the map $y_\alpha \to r_\alpha$ is a one-to-one map of Y into the rational numbers, and so Y is at most countable. The conclusion now follows easily since to each pair of rational numbers r, s with $y_\beta < r < s < x_\beta$ for some $\beta \in A$, one such interval $[y_\beta, x_\beta)$ can be associated, β denoted as $\beta_{r,s}$. Given $y \in \bigcup_{\alpha\in A} [y_\alpha, x_\alpha) - Y$, y lies in the interior of some interval $[y_\alpha, x_\alpha)$, so there exist r, s rational with $y_\alpha < r < s < x_\alpha$, and so y lies in one of the countably many intervals $[y_{\beta_{r,s}}, x_{\beta_{r,s}})$, and the proof is done. ∎

Proof of 1.1.18-1.1.20. The absolute continuity of fg and $1/g$ follows exactly as in the Lebesgue measure case. Using (1.1.17) then gives 1.1.18, 1.1.19, exactly as in basic calculus. 1.1.20 follows from 1.1.15 and 1.1.18. ∎

Proof of 1.1.21. Since f is bounded, it follows that F is uniformly continuous on the closure of the range of f, so that it is easily seen that $F \circ f$ is absolutely continuous with respect to ω. Using (1.1.17) again and proceeding as in the basic calculus proof of the chain rule, 1.1.21 follows. ∎

1.2.0 Weighted Hardy Inequalities

Tomaselli [TM], Talenti [TL], and Artola [AR] characterized the weights for which a Hardy inequality of type (1.2.10) or (1.2.12) with $p = q$ holds. A simpler proof was found by Muckenhoupt [M1], which in turn was generalized by Bradley [BR] to include the case $q > p$. The other inequalities dealt with in this section are not direct generalizations of the original Hardy inequalities but are similar in nature. Their importance stems from the fact that they arise naturally in the analysis of certain Sobolev inequalities.

It will be assumed that μ, λ are positive measues on $(\mathbf{R} \cup \{-\infty,\infty\},\mathbf{K})$, where $\mathbf{K}$ is the σ-algebra generated by the Borel sets $\mathbf{B}$ and the points $\{-\infty,\infty\}$; and ν is a positive measure on $(\mathbf{R},\mathbf{B})$ for which there is a Lebesgue decomposition with respect to Lebesgue measure. For notational simplicity, $(\int_A \bar{\nu}(t)^{-1/(p-1)}dt)^{p-1}$ will represent $\sup_A \bar{\nu}^{-1}$ when $p = 1$, $\bar{\nu}$ being the density of the absolutely continuous part of ν and $\frac{1}{p} + \frac{1}{p'} = 1$. The proofs of the following theorems will be deferred to 1.2.13.

1.2.1 Theorem. For $1 \leq p \leq q < \infty$,

$$(\int_{-\infty}^{\infty} (\int_t^{\infty} g(s)ds)^q d\mu(t))^{1/q} \leq c_1(\int_{-\infty}^{\infty} g^p(t)d\nu(t))^{1/p} \tag{1.2.2}$$

for some $c_1 > 0$ and all nonnegative Borel measurable g iff

$$\mu^{1/q}([-\infty,r]) \int_r^{\infty} g(s)ds \leq c_2(\int_r^{\infty} g^p(t)\bar{\nu}(t)dt)^{1/p} \tag{1.2.3}$$

for some $c_2 < \infty$ and all $r \in \mathbf{R}$ and g nonnegative and Borel measurable; iff

$$\mu^{1/q}([-\infty,r])(\int_r^{\infty} \bar{\nu}(t)^{-1/(p-1)}dt)^{1/p'} \leq c_3 \tag{1.2.4}$$

for some $c_3 < \infty$ and all $r \in \mathbf{R}$. And, by a reflection,

$$(\int_{-\infty}^{\infty} (\int_{-\infty}^{t} g(s)ds)^q d\mu(t))^{1/q} \leq c_1(\int_{-\infty}^{\infty} g^p(t)d\nu(t))^{1/p} \tag{1.2.5}$$

for some $c_1 < \infty$ and all nonnegative Borel measure g iff

$$\mu^{1/q}([r,\infty]) \int_{-\infty}^{r} g(s)ds \leq c_2(\int_{-\infty}^{\infty} g^p(t)\bar{\nu}(t)dt)^{1/p} \tag{1.2.6}$$

for some $c_2 < \infty$ and all $r \in \mathbf{R}$ and g nonnegative and Borel measurable; iff

$$\mu^{1/q}([r,\infty])(\int_{-\infty}^{r} \bar{\nu}(t)^{-1/(p-1)}dt)^{1/p'} \leq c_3 \tag{1.2.7}$$

for some $c_3 < \infty$ and all $r \in \mathbf{R}$, where $\bar{\nu}$ is the Lebesgue density of ν $(d\nu = \bar{\nu}(t)dt + d\nu_s)$.

If the constants c_i, $i = 1,2,3$ are chosen as small as possible, then $c_3 \leq c_2 \leq c_1 \leq p'^{1/p'}p^{1/q}c_3$. The convention $0\cdot\infty = 0$ is assumed, and $p'^{1/p'}$ is taken to be 1 when $p' = \infty$.

Given μ a positive measure on $(A,\mathbf{K})$ and ν a positive measure on (B,B) extend μ so that $\mu((\mathbf{R}\cup\{-\infty,\infty\}) - A) = 0$ and ν to have infinite density on $R - B$, then it easily follows from Theorem 1.2.1 that

$$\left(\int_A \left(\int_{B\cap(t,\infty)} g(s)ds\right)^q d\mu(t)\right)^{1/q} \le c\left(\int_B g(t)^p d\nu(t)\right)^{1/p} \tag{1.2.8}$$

for some $c < \infty$ and all nonnegative Borel measurable function g iff $\sup_r \mu^{1/q}([-\infty,r]\cap A)\left(\int_{B\cap(r,\infty)} \bar{\nu}(t)^{-1/(p-1)}dt\right)^{1/p'} < \infty$. Theorem 1.2.9 presents two special cases of this

<u>1.2.9</u> <u>Theorem</u>. For $1 \le p \le q < \infty$,

$$\left(\int_0^\infty \left(\int_t^\infty g(s)ds\right)^q d\mu(t)\right)^{1/q} \le c_1\left(\int_0^\infty g^p(t)d\nu(t)\right)^{1/p} \tag{1.2.10}$$

for some $c_1 < \infty$ and all nonnegative Borel measurable functions g, iff

$$\sup_{0<r} \mu^{1/q}([0,r])\left(\int_r^\infty \bar{\nu}^{-1/(p-1)}dt\right)^{1/p'} = b_1 < \infty . \tag{1.2.11}$$

And

$$\left(\int_0^\infty \left(\int_0^t g(s)ds\right)^q d\mu(t)\right)^{1/q} \le c_2\left(\int_0^\infty g^p(t)d\nu(t)\right)^{1/p} \tag{1.2.12}$$

for some c_2 and all nonnegative Borel measurable functions g iff

$$\sup_{0<r} \mu^{1/q}([r,\infty])\left(\int_0^r \bar{\nu}(t)^{-1/(p-1)}dt\right)^{1/p'} = b_2 < \infty . \tag{1.2.13}$$

And as a consequence,

$$\left(\int_{-\infty}^\infty \left|\int_0^t g(s)ds\right|^q d\mu(t)\right)^{1/q} \le c_3\left(\int_{-\infty}^\infty g^p(t)d\nu(t)\right)^{1/p} \tag{1.2.14}$$

for some $c_3 > 0$ and all nonnegative Borel measurable g iff

$$\begin{cases} \sup_{0<r} \mu^{1/q}([r,\infty])\left(\int_0^r \bar{\nu}(t)^{-1/(p-1)}dt\right)^{1/p'} = b_2 < \infty \\ \text{and} \\ \sup_{r<0} \mu^{1/q}([-\infty,r])\left(\int_r^0 \bar{\nu}(t)^{-1/(p-1)}dt\right)^{1/p'} = b_3 < \infty , \end{cases} \tag{1.2.15}$$

where $\bar{\nu}$ is as in Theorem 1.2.1.

If c_i, $i = 1,2,3$, b_i, $i = 1,2$ are chosen as small as possible, then $b_i \leq c_i \leq p^{1/q}p'^{1/p'}b_i$ for $i = 1,2$ and $\max\{b_2,b_3\} \leq c_3 \leq p^{1/q}p'^{1/p'}\max\{b_2,b_3\}$. The convention $\int_0^t g(s)ds = -\int_t^0 g(s)ds$ for $t < 0$ is used in (1.2.14) and $0\cdot\infty = 0$ is used throughout. Also $p'^{1/p'} = 1$ for $p' = \infty$.

The inequalities dealt with in Theorem 1.2.16 depart somewhat from the structure of the classical Hardy inequalities, but their analysis is similar. They arise naturally in the study of certain Sobolev inequalities. It is somewhat remarkable that (1.2.23) and (1.2.24) are equivalent since in general their left-hand sides are not comparable unless $|\lambda|$, $|\mu| < \infty$ and $\lambda(E) \leq c\mu(E)$.

<u>1.2.16 Theorem</u>. For $1 \leq p \leq q < \infty$,

$$\left(\int_{-\infty}^{\infty}\left(\int_t^{\infty}\int_t^{s} g(\sigma)d\sigma\, d\lambda(s)\right)^q d\mu(t)\right)^{1/q} \tag{1.2.17}$$

$$\leq c_{1,1}\left(\int_{-\infty}^{\infty} g^p(t)d\nu(t)\right)^{1/p}$$

for some $c_{1,1} < \infty$ and all nonnegative Borel measurable g iff

$$\mu^{1/q}([-\infty,r])\int_r^{\infty} g(s)\lambda[s,\infty]ds \leq c_{2,1}\left(\int_r^{\infty} g^p(t)d\nu(t)\right)^{1/p} \tag{1.2.18}$$

for some $c_{2,1} < \infty$ and all $r \in \mathbf{R}$ and g nonnegative and Borel measurable iff

$$\sup_r \mu^{1/q}([-\infty,r])\left(\int_r^{\infty}\left(\frac{\lambda^p[t,\infty]}{\bar{\nu}(t)}\right)^{1/(p-1)}dt\right)^{1/p'} = b_1 < \infty \tag{1.2.19}$$

And, by a reflection,

$$\left(\int_{-\infty}^{\infty}\left(\int_{-\infty}^{t}\int_s^{t} g(\sigma)d\sigma\, d\lambda(s)\right)^q d\mu(t)\right)^{1/q} \tag{1.2.20}$$

$$\leq c_{1,2}\left(\int_{-\infty}^{\infty} g^p(t)d\nu(t)\right)^{1/p}$$

for some $c_{1,2} < \infty$ and all nonnegative Borel measurable g iff

$$\mu^{1/q}([r,\infty])\int_{-\infty}^{r} g(s)\lambda[-\infty,r]ds \leq c_{2,2}\left(\int_{-\infty}^{r} g^p(t)d\nu(t)\right)^{1/p} \tag{1.2.21}$$

for some $c_{2,2} < \infty$ and all $r \in \mathbf{R}$ and g nonnegative and Borel measurable, iff

$$\sup_r \mu^{1/q}([r,\infty])\left(\int_{-\infty}^{r} \left(\frac{\lambda^p[-\infty,t]}{\bar{\nu}(t)}\right)^{1/(p-1)} dt\right)^{1/p'} = b_2 < \infty . \tag{1.2.22}$$

And, in consequence,

$$\left(\int_{-\infty}^{\infty} \left(\int_{-\infty}^{\infty} \left|\int_s^t g(\sigma)ds\right| d\lambda(s)\right)^q d\mu(t)\right)^{1/q} \tag{1.2.23}$$

$$\leq c_{1,3}\left(\int_{-\infty}^{\infty} g(t)d\nu(t)\right)^{1/p}$$

for some $c_{1,3} < \infty$ and all nonnegative Borel measurable g iff

$$\left(\int_{-\infty}^{\infty} \left|\int_{-\infty}^{\infty} \int_s^t g(\sigma)d\sigma \; d\lambda(s)\right|^q d\mu(t)\right)^{1/q} \tag{1.2.24}$$

$$\leq c_{2,3}\left(\int_{-\infty}^{\infty} g^p(t)d\nu(t)\right)^{1/p}$$

for some $c_{2,3} < \infty$ and all nonnegative Borel measurable g which are bounded and have compact support,
iff (1.2.18) and (1.2.21) hold,
iff (1.2.19) and (1.2.22) hold,
where $\bar{\nu}$ is as in Theorem 1.2.1.

The conventions $\int_s^t g(\sigma)d\sigma = -\int_t^s g(\sigma)d\sigma$ for $s > t$, $\frac{\lambda^p[t,\infty]}{\bar{\nu}(t)} = 0$ if the numerator and denominator are either both 0 or both ∞, and $0\cdot\infty = 0$ are used.

<u>1.1.25</u>. If the integral $\int_{-\infty}^{\infty} \int_s^t g(\sigma)d\sigma \; d\lambda(s)$ in (1.2.24) is not defined in the classical sense, that is, if t is given and $\int_s^t g(\sigma)d\sigma$ takes on both positive and negative values and is not in $L^1(\lambda)$, then it may be given an arbitrary value without affecting the theorem. Care must be taken if for fixed t, $\int_s^t g(\sigma)d\sigma$ is of one sign and is not in $L^1(\lambda)$, in which case the integral 1.2.25 is given the value ∞ or $-\infty$, depending on the sign of $\int_s^t g(\sigma)d\sigma$. The assumption that g is bounded and of compact support in (1.2.24) is added solely for use in the applications; it is not necessary here.

If $c_{i,j}$ are chosen as small as possible, then $b_i \leq c_{2,i} \leq c_{1,i} \leq p^{1/q}p'^{1/p'}b_i$ for $i = 1,2$ and $\max\{b_1,b_2\} \leq c_{2,3} \leq c_{1,3} \leq 2p^{1/q}p'^{1/p'}\max\{b_1,b_2\}$.

Remark. Theorems 1.2.1, 1.2.9, and 1.2.16 are equivalent. Theorems 1.2.9 and 1.2.16 will be proven directly from 1.2.1, and Theorem 1.2.1 may be recovered from 1.2.9 by a change of variable from $[0,\infty]$ to $[-\infty,\infty]$ accompanied by appropriate choices of measures, and from 1.2.16 by choosing λ to be a point mass at ∞ or $-\infty$.

It would be interesting to extend the preceding theorems to the case $q < p$. The following theorem extends the last part of Theorem 3 to the case $q = 1$. The global nature of condition renders it of no use in proving Sobolev inequalities.

1.2.26 Theorem. For $1 \le p < \infty$,

$$\int_{-\infty}^{\infty}\int_{-\infty}^{\infty}\left|\int_{s}^{t} g(\sigma)d\sigma\right| d\lambda(s)d\mu(t) \le c\left(\int_{-\infty}^{\infty} g^p(t)d\nu(t)\right)^{1/p} \tag{1.2.27}$$

for all nonnegative Borel measurable g iff

$$\left(\int_{-\infty}^{\infty}\left[\frac{(\lambda[-\infty,t)\mu(t,\infty]+\lambda(t,\infty]\mu[-\infty,t))^p}{\bar{\nu}(t)}\right]^{1/(p-1)} dt\right)^{1/p'} \le c , \tag{1.2.28}$$

$\bar{\nu}$ as in Theorem 1.2.1. The conventions $\frac{0}{0} = \frac{\infty}{\infty} = 0$ for the integrand of (1.2.28), $0\cdot\infty = 0$ and $\int_s^t g(\sigma)d\sigma = -\int_t^s g(\sigma)d\sigma$ for $s > t$ are used.

Proof of Theorem 1.2.1. The main substance of the result is the sufficiency of inequality (1.2.4). Assume (1.2.4) and $p > 1$. Let $h(t) = \left(\int_t^{\infty} \bar{\nu}(s)^{-1/(p-1)}ds\right)^{1/p'}$ and $I_\infty = \{t : \mu([-\infty,t]) = 0\}$ so that $\mu(I_\infty) = 0$. From (1.2.4) it follows that $h = \infty$ only on I_∞ and so $h < \infty$ on $T = \mathbf{R} - I_\infty$.

1.2.29. Let I_0 be the interval $I_0 = \{t : h(t) = 0\}$ so that h is locally absolutely continuous on $T - I_0$. This combined with the continuity of h on T leads to

$$h(t) = -\int_t^{\infty} h'(s)ds \quad \text{for} \quad t \in T . \tag{1.2.30}$$

1.2.31. If $g > 0$ and $\bar{\nu} = \infty$ on a set of positive measure, then (1.2.2) is true; otherwise $g = 0$ a.e. on $\{\bar{\nu} = \infty\}$, and so $g = p'^{1/p'}g(\bar{\nu}h)^{1/p}(h')^{1/p'}$ a.e. in T. Now using Hölder's inequality and (1.2.30), and recalling that $\mu(\mathbf{R} - T) = 0$, it follows that

$$\Big(\int_{-\infty}^{\infty}\Big(\int_t^{\infty} g(s)ds\Big)^q d\mu(t)\Big)^{p/q}$$

$$\leq p'^{p/p'}\Big(\int_{-\infty}^{\infty}\Big(\int_t^{\infty} g^p\bar{\nu}h\Big)^{q/p} h^{q/p'}(t)\,d\mu(t)\Big)^{p/q}$$

$$\leq p'^{p/p'}\int_{-\infty}^{\infty} g^p\bar{\nu}h\Big(\int_{-\infty}^{s} h^{q/p'}(t)d\mu(t)\Big)^{p/q}ds$$

(by using Minkowski's inequality)

$$\leq p'^{p/p'}c_3^{p-1}\int_{-\infty}^{\infty} g^p\bar{\nu}h\Big(\int_{[-\infty,s]\cap I'}\mu^{-1/p'}([-\infty,t])d\mu(t)\Big)^{p/q}ds$$

(by (1.2.4), where $I' = \{t : \mu[-\infty,t] = \infty\}$)

$$\leq p'^{p/p'}c_3^{p-1}p^{p/q}\int_{-\infty}^{\infty} g^p\bar{\nu}h\mu^{1/q}([-\infty,s] - I')ds$$

(using 1.2.32)

$$\leq p'^{p/p'}c_3^{p}p^{p/q}\int_{-\infty}^{\infty} g^p\bar{\nu}ds\ .$$

(by (1.2.4))

<u>1.2.32</u>. $[-\infty,s]\cap I'$ is an interval $[-\infty,s']$ or $(-\infty,s')$ on which $\mu([-\infty,t])$ is finite. Pick $s'' < s'$ and let $\bar{\omega}$ be the restriction of μ to $(-\infty,s'']$, and let ω be the reflection of $\bar{\omega}$, i.e. $\omega(A) = \omega(-A)$. Now apply the results of 1.1.13, specifically (1.1.22), with ω and $\lambda = \mu(\{-\infty\})$ to get that

$$\int_{[-\infty,s'']}\mu^{-1/p'}([-\infty,t])d\mu(t) \leq p\mu^{1/p}([-\infty,s''])\ . \tag{1.2.33}$$

Let $s'' \to s'$ and use $\mu^{-1/p'}(\{s'\})\mu(\{s'\}) = \mu^{1/p}(\{s'\})$ if s' is an atom of μ to get (1.2.33) with s'' replaced by s as required.

If $p = 1$, then

$$\Big(\int_{-\infty}^{\infty}\Big(\int_t^{\infty} g(s)ds\Big)^q d\mu(t)\Big)^{1/q}$$

$$\leq \int_{-\infty}^{\infty} g(s)\mu^{1/q}([-\infty,s])ds \quad \text{by Minkowski's inequality}$$

$$\leq c_3\int_{-\infty}^{\infty} g(s)\operatorname*{ess\,inf}_{(s,\infty)}\bar{\nu}\,ds \leq c_3\int_{-\infty}^{\infty} g(s)\bar{\nu}\,ds\ .$$

The fact that (1.2.2) implies (1.2.3) follows by first replacing g by $g\chi_{(r,\infty)\cap A}$, where $\mathbf{R} - A$ supports the singular part of ν and $|\mathbf{R} - A| = 0$ and then reducing the interval of integration with respect to μ to $[-\infty,\nu]$.

The proof of the implication (1.2.3) ==> (1.2.4) is broken down into three cases depending on whether $(\int_r^\infty \bar{\nu}(t)^{-1/(p-1)}dt)^{p-1}$ is zero, strictly positive but finite, or infinite (recall that for $p = 1$ this integral represents $\sup_{(r,\infty)} \bar{\nu}^{-1}$). In the first case (1.2.4) is trivial. In the second case if $p > 1$, set $g = \bar{\nu}^{-1/(p-1)}$, and if $p = 1$ set $g = g_n = \chi_{B_n}$, where $B_n = \{t : \bar{\nu}^{-1}(t) > -\frac{1}{n} + \sup_{(r,\infty)} \bar{\nu}^{-1}\}$, and let $n \to \infty$ to achieve (1.2.4).

1.2.34. In the third case it is necessary to construct a function g such that $g > 0$, $\int_r^\infty g^p\bar{\nu}\,dt < \infty$ while $\int_r^\infty g = \infty$, in which case, recalling the convention $0\cdot\infty = 0$, it is seen that $\mu([-\infty,r]) = 0$ and (1.2.4) is proven

To construct g as in 1.2.34 it is first assumed that $\bar{\nu} > 0$ a.e.; otherwise let g be ∞ on $\{\bar{\nu} = 0\}$ and zero elsewhere. For $p > 1$ let $E_n = \{t \in (r,\infty) : 2^{-(n+1)} < \bar{\nu}(t) < 2^{-n}\}$. If $|E_n| = \infty$ for some n, then pick g such that $g \in L^p(E_n)$, $g \notin L^1(E_n)$, and $g = 0$ elsewhere. Otherwise $|E_n| < \infty$ for all n.

$\sum_{n=-\infty}^{\infty} |E_n|2^{(n+1)/(p-1)} > \int_r^\infty \bar{\nu}^{-1/(p-1)} = \infty$, so pick i_k iteratively such that $i_0 = 0$ and for $n_k = \sum_{\ell=0}^{k-1} i_\ell$ it holds that

$$S_k = \sum_{n_k<|n|<n_k+i_k} |E_n|2^{(n+1)/(p-1)} > (k + 1)^\alpha$$

for a fixed $\alpha > \frac{2}{p - 1}$.

Let $g(t) = (\frac{2^{(1+pn)/(p-1)}}{(k + 1)^2 S_k})^{1/p}$ if $n_k < |n| < n_{k+1}$ and $t \in E_n$ so

$$\int_r^\infty g^p\bar{\nu} = \sum_{n=-\infty}^{\infty} \frac{|E_n|\ 2^{(n+1)/(p-1)}}{(k + 1)^2 S_k}$$

$$= \sum_{k=0}^{\infty} \frac{1}{(k+1)^2 S_k} \sum_{n_k \le |n| < n_{k+1}} |E_n| \, 2^{(n+1)/(p-1)}$$

$$= \sum_{k=0}^{\infty} \frac{1}{(k+1)^2} < \infty ,$$

while

$$\int_r^{\infty} g = 2^{-1/p} \sum_{n=-\infty}^{\infty} \frac{|E_n| \, 2^{(n+1)/(p-1)}}{((k+1)^2 S_k)^{1/p}}$$

$$= 2^{-1/p} \sum_{k=0}^{\infty} \left(\frac{1}{(k+1)^2 S_k}\right)^{1/p} \sum_{n_k \le |n| < n_{k+1}} |E_n| \, 2^{(n+1)/(p-1)}$$

$$= 2^{-1/p} \sum_{k=0}^{\infty} \frac{S_k^{1-1/p}}{(k+1)^{2/p}} \ge 2^{-1/p} \sum_{k=0}^{\infty} (k+1)^{\alpha(1-1/p)-2/p}$$

$$= \infty$$

since $\alpha\left(1 - \frac{1}{p}\right) - \frac{2}{p} > 0$.

For $p = 1$, $\inf\limits_{(r,\infty)} \bar{\nu} = 0$, so pick a set A of positive finite measure such that $\inf\limits_{A} \bar{\nu} = 0$. Either $\bar{\nu} = 0$ on a set B of positive measure in which case take $g = \infty$ on B and zero elsewhere, or else ε_n can be chosen such that $\varepsilon_n \downarrow 0$, $\frac{\varepsilon_{n+1}}{\varepsilon_n} < \frac{1}{2}$ and $|E_n| > 0$, where $E_n = \{t \in A : \varepsilon_{n+1} \le \bar{\nu}(t) < \varepsilon_n\}$. For this case let $a_n = |\{t \in A : \bar{\nu}(t) < \varepsilon_n\}|$ so that $|E_n| = a_n - a_{n+1}$, and let

$$g = \begin{cases} \dfrac{\varepsilon_n - \varepsilon_{n+1}}{a_n - a_{n+1}} \cdot \dfrac{1}{\varepsilon_n} & \text{on } E_n \\[2ex] 0 & \text{elsewhere} \end{cases}$$

It then follows that

$$\int_r^{\infty} g = \sum_{n=0}^{\infty} |E_n| \frac{\varepsilon_n - \varepsilon_{n+1}}{a_n - a_{n+1}} \cdot \frac{1}{\varepsilon_n}$$

$$= \sum_{n=0}^{\infty} \left(1 - \frac{\varepsilon_{n+1}}{\varepsilon_n}\right) \ge \sum_{n=0}^{\infty} \frac{1}{2} = \infty ,$$

while

$$\int_r^{\infty} g\bar{\nu} \le \sum_{n=0}^{\infty} |E_n| \frac{\varepsilon_n - \varepsilon_{n+1}}{a_n - a_{n+1}} \frac{\varepsilon_n}{\varepsilon_n} = \varepsilon_0 - \lim_{n\to\infty} \varepsilon_n = \varepsilon_0 < \infty .$$

The second half of the theorem is proven by replacing μ, ν, g by $\bar{\mu}$, $\bar{\nu}$, $\bar{g}$, where $\bar{\mu}(A) = \mu(-A)$, $\bar{\nu}(A) = \nu(-A)$, $\bar{g}(t) = g(-t)$, and using $\int_{-A} \bar{g}(t)d\bar{\mu}(t) = \int_A g(t)d\mu(t)$.

Proof of Theorem 1.2.9. Restrict μ to $A = [0,\infty]$ and ν to $B = [0,\infty)$, and then extend them as in (1.2.8). If (1.2.10) holds, then (1.2.2) holds with the extended measures since if $g > 0$ on a set of positive measure in $(-\infty,0)$, then the right-hand side of (1.2.2) is infinite. (1.2.11) then follows from (1.2.3). Conversely, if (1.2.11) is true, then (1.2.4) trivially holds for the extended measures, the condition for $r < \infty$ reducing to that of $r = 0$, and (1.2.10) follows from (1.2.2) by taking g with support in $[0,\infty)$. The equivalence of (1.2.12) and (1.2.13) follows similarly.

Assume (1.2.14). Letting g have support in $[0,\infty)$ and $(-\infty,0]$ respectively, it follows that (1.2.12) and its reflection

$$\left(\int_{-\infty}^{0} \left(\int_{t}^{0} g(s)ds\right)^q d\mu(t)\right)^{1/q} \le c_3\left(\int_{-\infty}^{0} g^p(t)d\nu(t)\right)^{1/p}$$

hold, which then implies (1.2.13) and its reflection, and so (1.2.15) holds.

Conversely, if (1.2.15) is true, then both (1.2.12) and its reflection hold so that

$$\begin{aligned}
&\int_{-\infty}^{\infty} \left|\int_{0}^{t} g(s)ds\right|^q d\mu(t) \\
&\quad = \int_{0}^{\infty} \left(\int_{0}^{t} g(s)ds\right)^q d\mu(t) + \int_{-\infty}^{0} \left(\int_{t}^{0} g(s)ds\right)^q d\mu(t) \\
&\quad \le \left(c_2\left(\int_{0}^{\infty} g^p(t)d\nu(t)\right)^{1/p}\right)^q + \left(c_4\left(\int_{-\infty}^{0} g^p(t)d\nu(t)\right)^{1/p}\right)^q \\
&\quad \le \max\{c_1^q, c_4^q\}\left(\int_{-\infty}^{\infty} g^p(t)d\nu(t)\right)^{q/p} . \quad \blacksquare
\end{aligned}$$

Proof of Theorem 1.2.16. The equivalence of (1.2.17), (1.2.18), and (1.2.19) will follow from that of (1.2.2), (1.2.3), and (1.2.4). The equivalence of (1.2.20), (1.2.21), and (1.2.22) then follows from applying the reflection $A \to -A$, as in Theorem 1.2.1.

It will now be shown that (1.2.17) $\Longrightarrow$ (1.2.18) $\Longrightarrow$ (1.2.19) $\Longrightarrow$ (1.2.17).

$$\int_t^\infty \int_t^s g(\sigma)d\sigma\, d\lambda(s) = \int_t^\infty g(\sigma)\lambda[\sigma,\infty]ds \quad \text{by Fubini.} \tag{1.2.35}$$

Assume (1.2.17), so

$$\left(\int_{-\infty}^\infty \left(\int_t^\infty g(s)\lambda[s,\infty]ds\right)^q d\mu(t)\right)^{1/q} \le c_{1,1}\left(\int_{-\infty}^\infty g^p(t)d\nu(t)\right)^{1/p}.$$

Replace g by $g\cdot\chi_{(r,\infty)}$ and reduce the interval of integration on the left to get $\mu^{1/q}[-\infty,r]\int_r^\infty g(s)\lambda[s,\infty]ds \le c_{1,1}\left(\int_{-\infty}^\infty g^p(t)d\nu(t)\right)^{1/p}$, and (1.2.18) is verified.

Assume (1.2.18), replace g by $g\cdot\chi_A$, where A is the support of the singular part of ν, to get

$$\mu^{1/q}[-\infty,r]\int_r^\infty g(s)\lambda[s,\infty]ds \le c_{2,1}\left(\int_{-\infty}^\infty g^p(t)\bar{\nu}(t)dt\right)^{1/p}. \tag{1.2.36}$$

Let $I_0 = \{t : \mu[-\infty,t] = 0\}$ so $\mu(I_0) = 0$ since I_0 is an interval. Let $J_0 = \{t : \lambda[t,\infty] = 0\}$ and $J_\infty = \{t : \lambda[t,\infty] = \infty\}$. From (1.2.36) it is seen that $\bar{\nu} = \infty$ a.e. on the interval $J_\infty - I_0$ (let $g = g_n = \chi_{\{\bar{\nu}<n\}\,(J_\infty - I_0)\,(-\infty,n)}$), so if $\bar{\nu}_*(t) = \dfrac{\bar{\nu}(t)}{\lambda^p[t,\infty]}$, then $\bar{\nu}_*(t) = \infty$ on $J_\infty - I_0$ using the convention that $\dfrac{\lambda^p[t,\infty]}{\bar{\nu}(t)} = 0$ if $\lambda[t,\infty]$ and $\bar{\nu}(t)$ are either both 0 or both ∞. Also $\bar{\nu}_*(t) = \infty$ on J_0 using the same convention.

If it doesn't hold that $g = 0$ a.e. on $(J_\infty - I_0)\cup J_0$, then

$$\mu^{1/q}[-\infty,r]\int_r^\infty g(s)ds \le c_{2,1}\int_r^\infty g^p(t)\bar{\nu}_*(t)dt, \tag{1.2.37}$$

otherwise let $\bar{g}(t) = \begin{cases} 0, & J_0 \cup J_\infty \\ \dfrac{g(t)}{\lambda[t,\infty]}, & \text{otherwise} \end{cases}$, so on $\mathbf{R} - I_0$, $g(t) = \bar{g}(t)\lambda[t,\infty]$ and $\bar{g}^p\bar{\nu} = g^p\bar{\nu}_*$. Using $\bar{g}$ in (1.2.36) then gives (1.2.37), but since (1.2.3) $\Longrightarrow$ (1.2.4), it follows that $\mu^{1/q}[-\infty,r]\left(\int_r^\infty \bar{\nu}_*^{-1/(p-1)}(t)dt\right)^{1/p'} \le c_{2,1}$, so (1.2.19) is verified.

Assume (1.2.19). Using (1.2.4) $\Longrightarrow$ (1.2.2), it follows that

$$\left(\int_{-\infty}^\infty \left(\int_t^\infty g(s)ds\right)^q d\mu(t)\right)^{1/q} \le p^{1/q}p'^{1/p'}b_1\left(\int_{-\infty}^\infty g^p(t)\bar{\nu}_*(t)dt\right)^{1/p}$$

Replace $g(s)$ by $g(s)\lambda[s,\infty]$ and use that $\lambda[t,\infty]\bar{\nu}_*(t) \le \bar{\nu}(t)$ to get (1.2.17) and so the circle of implications is completed.

It remains to show that (1.2.23) $\Longrightarrow$ (1.2.24) $\Longrightarrow$ {(1.2.18), (1.2.21)} $\Longrightarrow$ {(1.2.19), (1.2.22)} $\Longrightarrow$ (1.2.23). Recall 1.2.25. (1.2.23) $\Longrightarrow$ (1.2.24) is trivial. To show (1.2.24) $\Longrightarrow$ (1.2.18), first reduce the interval of integration on the far left of (1.2.24) to $[-\infty,r]$ and replace g by $g\chi_{(r,\infty)}$ to get

$$\mu^{1/q}[-\infty,r]\int_r^\infty g(s)\lambda[s,\infty]ds$$

$$= \Big(\int_{-\infty}^r \Big|\int_r^\infty \int_r^s g(\sigma)d\sigma\, d\lambda(s)\Big|^q d\mu(t)\Big)^{1/q} \le c_{2,3}\Big(\int_r^\infty g^p(t)d\nu(t)\Big)^{1/p}$$

for bounded g of compact support since $\int_s^t g(\sigma)\chi_{[r,\infty)}d\sigma = 0$ if both $s,\ t \le r$. Take monotone limits to get all positive measurable g. In a similar way (1.2.24) $\Longrightarrow$ (1.2.21). {(1.2.18), (1.2.21)} $\Longrightarrow$ {(1.2.19), (1.2.22)} by the first part of Theorem 1.2.16. From Theorem 1.2.1 it is seen that {(1.2.19), (1.2.22)} $\Longrightarrow$ {(1.2.2), (1.2.5)}, with $\bar\nu(t)$ replaced by $\frac{\bar\nu(t)}{\lambda[t,\infty]}$ so replacing $g(t)$ by $g(t)\lambda[t,\infty]$ and using that $\lambda[t,\infty]\bar\nu_*(t) \le \bar\nu(t)$, it follows that

$$\Big(\int_{-\infty}^\infty \Big(\int_{-\infty}^\infty \Big|\int_s^t g(\sigma)d\sigma\Big| d\lambda(s)\Big)^q d\mu(t)\Big)^{1/q}$$

$$\le \Big(\int_{-\infty}^\infty \Big(\int_t^\infty \int_t^s g(\sigma)d\sigma\, d\lambda(s)\Big)^q d\mu(t)\Big)^{1/q}$$

$$+ \Big(\int_{-\infty}^\infty \Big(\int_{-\infty}^t \int_s^t g(\sigma)d\sigma\, d\lambda(s)\Big)^q d\mu(t)\Big)^{1/q}$$

$$= \Big(\int_{-\infty}^\infty \Big(\int_t^\infty g(\sigma)\lambda[\sigma,\infty]d\sigma\Big)^q d\mu(t)\Big)^{1/q}$$

$$+ \Big(\int_{-\infty}^\infty \Big(\int_{-\infty}^t g(\sigma)\lambda[-\infty,\sigma]d\sigma\Big)^q d\mu(t)\Big)^{1/q}$$

$$\le 2p^{1/q}p'^{1/p'}\max\{b_1,b_2\}\Big(\int_{-\infty}^\infty g^p(t)\bar\nu(t)dt\Big)^{1/p},$$

and the proof is complete.

Proof of Theorem 1.2.26.

$$\int_{-\infty}^{\infty}\int_{-\infty}^{\infty}\left|\int_{s}^{t} g(\sigma)d\sigma\right|d\lambda(s)d\mu(t)$$

$$= \int_{-\infty}^{\infty}\left(\int_{t}^{\infty} g(\sigma)\lambda(\sigma,\infty]d\sigma + \int_{-\infty}^{t} g(\sigma)\lambda[-\infty,\sigma)d\sigma\right)d\mu(t)$$

$$= \int_{-\infty}^{\infty} g(\sigma)\int_{-\infty}^{\infty} f(\sigma,t)d\mu(t)d\sigma$$

$$\text{for}\quad f(\sigma,t) = \begin{cases} \lambda(\sigma,\infty]\ , & \sigma > t \\ 0\ , & \sigma = t \\ \lambda[-\infty,\sigma), & \sigma < t \end{cases}$$

$$= \int_{-\infty}^{\infty} g(\sigma)h(\sigma)d\sigma \tag{1.2.38}$$

for $h(\sigma) = \lambda(\sigma,\infty]\mu[-\infty,\sigma) + \lambda[-\infty,\sigma)\mu(\sigma,\infty]$.

Let $E_0 = \{t : h(t) = 0\}$, $E_\infty = \{t : h(t) = \infty\}$.

Assume (1.2.27). Replace g by $g\chi_A$, where A is the support of the singular part of ν to get

$$\int_{-\infty}^{\infty} g(\sigma)h(\sigma)d\sigma \le c\left(\int_{-\infty}^{\infty} g^p(t)\bar{\nu}(t)dt\right)^{1/p} . \tag{1.2.39}$$

From this it is easy to see that $\bar{\nu} = \infty$ on E_∞ so that $\bar{\nu}_*(t) = \dfrac{\bar{\nu}(t)}{h^p(t)} = \infty$ on E_∞ using the convention for $\dfrac{h^p(t)}{\bar{\nu}(t)}$ that $\frac{\infty}{\infty} = 0$. Also $\bar{\nu}_*(t) = \infty$ on E_0 using the $\frac{0}{0} = 0$ convention for $\dfrac{h^p(t)}{\bar{\nu}(t)}$. If it is not true that $g = 0$ a.e. on $E_0 \cup E_\infty$, then

$$\int_{-\infty}^{\infty} g(s)ds \le c\left(\int_{-\infty}^{\infty} g^p(t)\bar{\nu}_*(t)dt\right)^{1/p} , \tag{1.2.40}$$

otherwise let $\bar{g}(t) = \begin{cases} 0 & E_0 \cup E_\infty \\ \dfrac{g(t)}{h(t)}, & \text{elsewhere} \end{cases}$, and substitute $\bar{g}$ in (1.2.39) to get (1.2.40). Assuming $\left(\int_{-\infty}^{\infty} \bar{\nu}_*(t)^{-1/(p-1)}dt\right)^{p-1} = \infty$ leads to a contradiction with (1.2.40) using the construction in Theorem 1 since either $\left(\int_0^{\infty} \bar{\nu}_*(t)^{-1/(p-1)}dt\right)^{p-1} = \infty$ or $\left(\int_{-\infty}^{0} \bar{\nu}_*(t)^{-1/(p-1)}dt\right)^{p-1} = \infty$. If $\left(\int_{-\infty}^{\infty} \bar{\nu}_*(t)^{-1/(p-1)}dt\right)^{p-1} = 0$, then (1.2.28 is trivial; otherwise for $p > 1$ let $g(t) = \bar{\nu}_*(t)^{-1/(p-1)}$ in (1.2.40) to get (1.2.28). For $p = 1$ let $g_n(t) = \frac{1}{n}\chi_{[t-n,t+n]}$

in (1.2.40), and let $n \to \infty$ to differentiate the integrals and achieve (1.2.28).

Assume (1.2.28). The integrability of $(\frac{h^p}{\bar{\nu}})^{1/(p-1)}$ implies that $\frac{h^p}{\bar{\nu}} < \infty$ a.e., and so considering the $\frac{0}{0} = 0$ convention, off a set of positive measure

$$\text{either } \bar{\nu}(t) = 0 \text{ or } h(t) = 0 \text{ for a given } t . \tag{1.2.41}$$

Also it can be assumed that $g = 0$ a.e. on $\{t : \bar{\nu}(t) = \infty\}$ since otherwise (1.2.27) is trivially true. Considering this and (1.2.41) it follows that $g(t)h(t) \leq g(t)\bar{\nu}^{1/p}(t)(\frac{h^p(t)}{\bar{\nu}(t)})^{1/p}$ a.e. and so Hölder's inequality applied to (1.2.38) gives (1.2.27).

1.3.0 Equivalence of Capacities

The set functions which arise naturally in the analysis of the Sobolev inequalities treated in Chapter 2 are difficult to work with in their original form except in special cases. In the present section they are shown to be comparable to more familiar capacities and set functions.

Let $(M,\mathbf{F},\nu)$ be a measure space with ν positive, and let $W_0^{1,p}(\nu,M)$ be a set of real valued $\mathbf{F}$ measurable functions on M satisfying the following properties. $W_0^{1,p}(\nu,M)$ is closed under composition with functions $f \in N = \{f \in C^\infty(\mathbf{R}) : f(0) = 0,\ f'$ is bounded, and $f' \geq 0\}$. There is a map $|D| : W_0^{1,p}(\nu,M) \to L^p(\nu,M)$ such that

$$(\int |Df \circ \phi|^p d\nu)^{1/p} \sim_d (\int |f'(\phi)|^p |D\phi|^p d\nu)^{1/p} \tag{1.3.1}$$

for all $\phi \in W_0^{1,p}(\nu,M)$, where the notation $|D\phi| = |D|(\phi)$ and $a \sim_d b$ iff $d^{-1}a \leq b \leq da$ has been used.

If $\mathbf{H} \subseteq W_0^{1,p}(\nu,M)$, $\mathbf{H}$ closed under composition with $f \in N$ and $1 \leq p < \infty$, $A \subseteq M$, then let

$$C_{\mathbf{H},p}(A) = \inf\{\int |D\phi|^p d\nu : \phi \in \mathbf{H},\ \phi \geq 1 \text{ on } A\} ,$$

$$K_{\mathbf{H},p}(A) = \inf\{(\int_0^1 \bar{\mu}_\phi^*(t)^{-1/(p-1)} dt)^{-(p-1)} : \phi \in \mathbf{H},\ \phi \geq 1 \text{ on } A\} .$$

$\bar{\mu}_\phi^*$ is the density of the absolutely continuous part of μ_ϕ^*, the distribution measure of μ_ϕ with respect to ϕ, where $d\mu_\phi = |D\phi|^p d\nu$, i.e., $\mu_\phi^*(E) = \int_{\phi^{-1}(E)} |D\phi|^p d\nu$. The conventions are used that $(\int_0^1 \bar{\mu}_\phi^*(t)^{-1/(p-1)} dt)^{-(p-1)} = \inf_{(0,1)} \bar{\mu}_\phi^*(t)$ if $p = 1$, and $\inf_{\mathbf{G}} = \infty$ if $\mathbf{G} = \emptyset$.

1.3.2 Theorem. $K_{\mathbf{H},p}(A) \sim_{d^p} C_{\mathbf{H},p}(A)$ for $A \subseteq M$.

Proof. Let $\mathbf{H}_A = \{\phi \in \mathbf{H} : \phi \geq 1 \text{ on } A\}$, so $\mathbf{H}_A$ is closed under composition with $f \in N^* = \{f \in N : f(1) = 1\}$.

$$\bar{\mu}^*_\phi \in L^1(\mathbf{R}) \,, \tag{1.3.3}$$

since $\int_{\mathbf{R}} \bar{\mu}^*_\phi(t)dt \leq \int_{\mathbf{R}} d\bar{\mu}_\phi = \int |D\phi|^p d\nu$, which is finite since $\phi \in W_0^{1,p}(\nu,M)$. Given $f \in N^*$, let $g = f'$; so $\int_0^1 g(t)dt = 1$ and $\int_{\mathbf{R}} g^p d\mu^*_\phi = \int g^p(\phi)|D\phi|^p d\nu$. Use (1.3.1) and lemma 1 with μ^*_ϕ as the measure on $\mathbf{R}$, $a = 0$, $b = 1$, $\sigma = 1$, and $I = \mathbf{R}$, to get $\inf_{N^*} \int |Df \circ \phi|^p d\nu \sim_{d^p} (\int_0^1 \bar{\mu}^*_\phi(t)^{-1/(p-1)}dt)^{-(p-1)}$. Taking the infimum over $\mathbf{H}_A$ gives $K_{\mathbf{H},p}(A) \sim_{d^p} C_{\mathbf{H},p}(A)$ since the function $f(x) = x$ is in N^* and $\mathbf{H}_A$ is closed under composition with functions in N^*. ∎

Let $(M,\mathbf{F},\lambda)$ be a measure space and $W^{1,p}(\nu,M)$ be a set of real-valued $\mathbf{F}$ measurable functions closed under composition with $f \in N' = \{f \in C^\infty : f' \text{ is bounded and of one sign}\}$ and on which $|D|$ is defined as before. If $\mathbf{H} \subseteq W^{1,p}(\nu,M)$, $\mathbf{H}$ closed under composition with $f \in N'$, $1 \leq p < \infty$ and $A \subseteq M$, then let

$$\bar{C}_{\mathbf{H},p}(A) = \inf\{\int |D\phi|^p d\nu : \phi \in \mathbf{H} \cap L^1(\lambda,M) \,,\ \phi \geq 1 \text{ on } A,\ \int \phi \, d\lambda = 0\} \,,$$

$$\bar{C}'_{\mathbf{H},p}(A) = \inf\{\int |D\phi|^p d\nu : \phi \in \mathbf{H} \cap L^1(\lambda,M) \,,\ \phi \leq 0 \text{ on } A,\ \int \phi \, d\lambda = 1\} \,,$$

$$\bar{K}_{\mathbf{H},p}(A) = \inf\{(\int_0^\infty (\frac{\lambda^p(\{\phi > t\})}{\bar{\mu}^*_\phi(t)})^{1/(p-1)} dt)^{-(p-1} : \phi \in \mathbf{H},\ \phi \leq 0 \text{ on } A\} \,.$$

The conventions $(\int_0^\infty (\frac{\lambda^p(\{\phi > t\})}{\bar{\mu}^*_\phi(t)})^{1/(p-1)} dt)^{-(p-1)} = \inf_{(0,\infty)} \frac{\bar{\mu}^*_\phi(t)}{\lambda^p(\{\phi > t\})}$ if $p = 1$, $\inf_{\mathbf{G}} = \infty$ if $\mathbf{G} = \emptyset$ and $\frac{0}{0} = 0$, $\frac{\infty}{\infty} = 0$ for the ratio $\frac{\lambda^p(\{\phi > t\})}{\bar{\mu}^*_\phi(t)}$ are used.

All of the expressions above are comparable (equal if $d = 1$, except for one pathological case, this being if

$\exists\ \phi \in \mathbf{H},\ \phi \leq 0$ on A such that (1.3.4)

$\lambda(\{\phi > t\}) = \infty$ for some $t > 0$.

It is clear that this is impossible if $\lambda(M) < \infty$. In applications $\lambda(M)$ is typically equal to one.

1.3.5 Theorem.

$$\bar{K}_{\mathbf{H},p}(A) \sim_{d^p} \bar{C}'_{\mathbf{H},p}(A) \quad \text{if (1.3.4) does not hold ,} \tag{1.3.6}$$

$$\bar{C}_{\mathbf{H},p}(A) \sim_{d^p} \bar{C}'_{\mathbf{H},p}(A) \quad \text{if } \lambda(M) = 1 , \tag{1.3.7}$$

$$\text{If (1.3.4) holds, then } \bar{K}_{\mathbf{H},p}(A) = 0 . \tag{1.3.8}$$

Remark. Under fairly general circumstances, it is possible to show that another comparable expression is

$$\inf\Big\{\int \frac{|D\phi|^p}{\lambda^p(\{\phi > \phi(x)\})}\, d\nu(x) : \phi \in \mathbf{H},\quad \phi \leq 0 \text{ on } A ,$$

$$\inf\{t : \lambda(\{\phi > t\}) = 0\} = 1\Big\} .$$

Proof of Theorem 1.3.5. (1.3.8) follows from the definition of $\bar{K}_{\mathbf{H},p}$ and (1.3.3). (1.3.7) follows by replacing ϕ with $1 - \phi$ and using $f(x) = 1 - x$ in (1.3.1).

1.3.9. It can be assumed that there exists a $\phi \in \mathbf{H}$ with $\phi \leq 0$ on A and $\lambda(\{\phi > 0\}) > 0$ since otherwise $\bar{K}_{\mathbf{H},p}(A) = \infty$ from its definition and the $\frac{0}{0} = 0$ convention, and $\bar{C}_{\mathbf{H},p}(A) = \infty$ using the convention $\inf_{\mathbf{G}} = \infty$ if $\mathbf{G} = \emptyset$, since it would be true that $\int \phi\, d\lambda \leq 0$.

1.3.10. In addition to 1.3.9, assume that (1.3.4) does not hold.

Given $\phi \in \mathbf{H}$ with $\phi \leq 0$ on A and $b = \inf\{t \in [0,\infty) : \lambda(\{\phi > t\}) = 0\}$ ($b = \infty$ possible), then $b > 0$ by 1.3.9, and $0 < \lambda(\{\phi > t\}) < \infty$ by 1.3.10. If $d\nu' = \dfrac{d\mu_\phi^*}{\lambda^p(\{\phi > t\})}$, then $\nu'(K) < \infty$ and $0 < C_k \leq \lambda(\{\phi > t\}) \leq C'_k < \infty$ for some C_k, C'_k if $K \subseteq (0,b)$, K compact and $t \in K$ since $\lambda(\{\phi > t\})$ is monotone and $\mu_\phi^*(\mathbf{R}) < \infty$. Applying Lemma 1.3.14 with $I = (0,b)$, $a = 0$, and $\sigma(t) = \lambda(\{\phi > t\})$, it follows that

$$\left\{\begin{aligned}&\left(\int_0^b \left(\frac{\lambda^p(\{\phi > t\})}{\bar{\mu}_\phi^*(t)}\right)^{1/(p-1)} dt\right)^{-(p-1)}\\ &\quad = \inf\{\int_{(0,b)} g^p(t)\,d\mu_\phi^*(t) : g \in \mathbf{F},\ \ g > 0,\\ &\qquad \text{and } \int_0^b g(t)\lambda(\{\phi > t\})dt = 1\}\end{aligned}\right. \tag{1.3.11}$$

for both $\mathbf{F} = C^\infty(\mathbf{R}) \cap L^\infty(\mathbf{R})$ and $\mathbf{F} = C_0^\infty(0,b)$. Let

$$\begin{aligned}\mathbf{L} = \{g \in C^\infty(\mathbf{R}) \cap L^\infty(\mathbf{R}) : g > 0\ ,\\ \int_0^\infty g(t)\lambda(\{\phi > t\})dt + \int_{-\infty}^0 g(t)\lambda(\{\phi < t\})dt < \infty, \quad \text{and}\\ \int_0^\infty g(t)\lambda(\{\phi > t\})dt - \int_{-\infty}^0 g(t)\lambda(\{\phi < t\})dt = 1\ .\end{aligned}$$

Considering all $g^* = g(\int_0^b g(t)\lambda(\{\phi > t\})dt)^{-1}$ for $g \in \mathbf{L}$ and noticing that $\int_0^b g(t)\lambda(\{\phi > t\})dt = \int_0^\infty g(t)\lambda(\{\phi > t\})dt > 1$ for $g \in \mathbf{L}$ by the definition of b and $\mathbf{L}$, it follows that (1.3.11) with $\mathbf{F} = C^\infty(\mathbf{R}) \cap L^\infty(\mathbf{R})$ is no larger than

$$\inf_{g\in\mathbf{L}} \int g^p(t)\,d\mu_\phi^*(t)\ . \tag{1.3.12}$$

But $\mathbf{L}$ contains $\mathbf{F} = C_0^\infty(0,b)$, so the opposite inequality is true, therefore (1.3.11) and (1.3.12) are equal.

Now let $f(t) = \int_0^t g(s)ds$ using the convention that $\int_t^0 = -\int_0^t$ so that $\int_0^\infty g(t)\lambda(\{\phi > t\})dt - \int_{-\infty}^0 g(t)\lambda(\{\phi < t\})dt = \int f(\phi)d\lambda$. This holds because $\int_0^\infty f'(t)\lambda(\{\phi > t\})dt = \int_0^\infty f'(t)\lambda(\{f(\phi) > f(t)\})dt$ since $f'(t) = 0$, where f is not one to one, and so a change of variables gives

$$\begin{aligned}\int_0^\infty f'(t)\lambda(\{\phi > t\})dt &= \int_{f(0)}^{f(\infty)} \lambda(\{\phi > t\})dt\\ &= \int_{f(0)}^\infty \lambda(\{f(\phi) > t\})dt = \int_{\{f(\phi)>0\}} f(\phi)d\lambda\end{aligned}$$

since $f(0) = 0$, and then a similar calculation handles the other integral. It is now clear that

$$\begin{cases} \inf\limits_{g\in L} \int g^p(t)\,d\mu_\phi^*(t) = \inf\{\int f'(\phi)^p |D\phi|^p d\nu : f \in C^\infty, \\ \qquad f' > 0, \quad f' \text{ bounded}, \quad f(0) = 0, \\ \qquad f(\phi) \in L^1(\lambda,M), \quad \text{and} \quad \int f(\phi)\,d\lambda = 1\}\ . \end{cases} \tag{1.3.13}$$

Finally, using (1.3.1) and taking the infimum over $\phi \in H$ with $\phi < 0$ on A and $\int \phi\, d\lambda = 1$, it follows that $\bar{K}_{H,p}(A) \sim_{d^p} \bar{C}'_{H,p}(A)$, since H is closed under composition with f of the type described in (1.3.13), one of which is $f(x) = x$. ∎

1.3.14 Lemma. Suppose ν is a positive Borel measure, $\sigma : \mathbf{R} \to \mathbf{R}$ is a Borel measurable function, $a,b \in \mathbf{R} \cup \{-\infty,\infty\}$, $a < b$, and I is an interval, possibly unbounded, such that $\nu(K) < \infty$ and $0 < C_K < \sigma(x) < C'_K < \infty$ for K compact, $x \in K$ and $K \subseteq (a,b) \subseteq I$, then

$$\inf_{g\in F} \int_I g^p(t)\,d\nu(t) = \left(\int_a^b \left(\frac{\sigma^p(t)}{\bar{\nu}(t)}\right)^{1/(p-1)} dt\right)^{-(p-1)}, \tag{1.3.15}$$

where F may be any subcollection of $G = \{g : \mathbf{R} \to \mathbf{R} : g$ is Borel measurable, $g > 0$, $\int_a^b g(t)\sigma(t)dt = 1\}$ which contains $G \cap C_0^\infty(a,b)$. $\bar{\nu}$ is the density of the absolutely continuous part of ν. The convention $\left(\int_a^b \left(\frac{\sigma^p(t)}{\bar{\nu}(t)}\right)^{1/(p-1)} dt\right)^{-(p-1)} = \inf\limits_{(a,b)} \frac{\bar{\nu}}{\sigma}$ will be used if $p = 1$.

Proof. (1.3.15) will first be proven for $d\nu(t) = \bar{\nu}(t)dt$ both with $F = G$ and $F = G \cap C_0^\infty(a,b)$. Then Lemma 1.3.16 will imply (1.3.15) in the general case for $F = G \cap C_0^\infty(a,b)$. It then follows that (1.3.15) is true for all intermediate subcollections of G.

Assume $d\nu = \bar{\nu}(t)dt$. Once (1.3.15) is proven for $F = G$, a smoothing argument will be given to prove (1.3.15) for $F = G \cap C_0^\infty(a,b)$.

If $p = 1$, then for $x \in (a,b)$ let $g_n = \chi_{I_n} \left(\int_{I_n} \sigma(t)dt\right)^{-1}$, $I_n = [x - \frac{1}{n}, x + \frac{1}{n}]$, which is defined for large n since then $I_n \subset (a,b)$. It is now seen that $\int_a^b g_n(t)\sigma(t)dt = 1$ and

$$\begin{aligned} \int_I g_n^p(t)\,d\nu(t) &= \int_I g_n^p(t)\bar{\nu}(t)dt \\ &= \left(2n \int_{I_n} \bar{\nu}(t)dt\right)\left(2n \int_{I_n} \sigma(t)dt\right)^{-1} \to \frac{\bar{\nu}(x)}{\sigma(x)} \end{aligned}$$

for almost all $x \in (a,b)$. Therefore

$$\inf_{g \in G} \int_I g^p(t)\,d\nu(t) \le \inf_{(a,b)} \frac{\bar{\nu}}{\sigma}.$$

In addition,

$$\int_I g^p(t)\,d\nu(t) = \int_I g^p(t)\bar{\nu}(t)\,dt$$

$$\ge \int_a^b g^p(t)\sigma(t)\,dt \inf_{(a,b)} \frac{\bar{\nu}}{\sigma} = \inf_{(a,b)} \frac{\bar{\nu}}{\sigma}.$$

If $p > 1$, then (1.3.15) will be proven for $\sigma = 1$, in which case making the substitutions $g = g'\sigma$ and $\bar{\nu} = \frac{\bar{\nu}'}{\sigma}$ and recalling that $0 < \sigma < \infty$ on (a,b), it follows that (1.3.15) holds for general σ.

Assuming $\sigma = 1$, an inequality in one direction is obtained by letting $g = \chi_{(a,b)}\bar{\nu}^{-1/(p-1)}\left(\int_a^b \bar{\nu}(t)^{-1/(p-1)}dt\right)^{-1}$ as long as $\int_a^b \bar{\nu}(t)^{-1/(p-1)}dt < \infty$, otherwise a construction virtually identical to that in the proof of the first Hardy inequality gives g such that $\int_a^b g(t)\,dt = \infty$ and $\int_a^b g^p(t)\bar{\nu}(t)\,dt < \infty$ and so, letting $a_n = \max\{a,-n\}$, $b_n = \min\{b,n\}$,

$$E_n = \{g \le n\} \cap (a_n,b_n)$$

and

$$g_n = g\chi_{E_n}\left(\int_a^b g\chi_{E_n}\,dt\right)^{-1},$$

so that $\int_a^b g_n\,dt = 1$ and

$$\int_a^b g_n^p\bar{\nu}\,dt \le \left(\int_a^b g^p\bar{\nu}\,dt\right)\left(\int_a^b g\chi_{E_n}\,dt\right)^{-p} \to 0 \quad \text{as} \quad n \to \infty,$$

the same inequality follows.

The opposite inequality is a consequence of Jensen's inequality. The inf is not increased if only $g \in G$ supported in (a,b) are considered. Given such a g, let g_n be as above.

$$\left(\int_a^b (\bar{\nu} + \varepsilon)^{-1/(p-1)}dt\right)^{-(p-1)}$$

$$\le \left(\int_{\operatorname{spt} g_n} (\bar{\nu} + \varepsilon)^{-1/(p-1)} g_n^{-1}\right) g_n\,dt\Big)^{-(p-1)} \le \int_a^b g_n^p(\bar{\nu} + \varepsilon)\,dt$$

by Jensen's inequality since $\int_a^b g_n dt = 1$.

Let $\varepsilon \to 0$ using the monotone convergence theorem on the left, then

$$\int_a^b g_n^p \bar{\nu} dt = \frac{\int_a^b g^p \chi_{E_n} \bar{\nu} dt}{(\int_a^b g\chi_{E_n} dt)^p} \to \int_a^b g^p \bar{\nu} dt$$

as well by the monotone convergence theorem, and so the opposite inequality holds and therefore equality as well.

(1.3.15) will now be proven for smooth g. Given $g \in G$, pick g_n bounded and positive with compact support in (a,b) such that $g_n \uparrow g$ in (a,b). Let δ_m be a C^∞ approximate identity with $\delta_m > 0$, $\int \delta_m = 1$, and the diameter of the support of $\delta_m \to 0$. Thus $\delta_m * g_n$ has compact support in (a,b) for large m and is bounded independent of m, so

$$\lim_{m\to\infty} \int_a^b g_n * \delta_m \sigma dt = \int_a^b g_n \sigma dt$$

and

$$\lim_{m\to\infty} \int_a^b (g_n * \delta_m)^p \bar{\nu} dt = \int_a^b g_n^p \bar{\nu} dt$$

by the dominated convergence theorem since $\bar{\nu}$ is integrable on compact subsets of (a,b) and σ is bounded uniformly away from $0,\infty$ on the support of $g_n * \delta_m$.

The monotone convergence theorem now leads to

$$\lim_{n\to\infty} \lim_{m\to\infty} \int_a^b g_n * \delta_m = \int_a^b g\sigma = 1$$

and

$$\lim_{n\to\infty} \lim_{m\to\infty} \int_a^b (g_n * \delta_m)^p \bar{\nu} dt = \int_a^b g^p \bar{\nu} dt .$$

From this can be extracted a sequence $\{f_k\}$, $f_k = g_{n_k} * \delta_{m_k}$ such that $f_k \in C_0^\infty(a,b)$, $\int_a^b f_k \sigma \to 1$, and $\int_a^b f_k^p \bar{\nu} dt \to \int_a^b g^p \bar{\nu} dt$. Letting $\bar{f}_k = \frac{f_k}{\int_a^b f_k \sigma dt}$, if follows that $\int_a^b \bar{f}_k \sigma dt = 1$ and $\int_a^b f_k^p \bar{\nu} dt \to \int_a^b g^p \bar{\nu} dt$ and 1.3.14 is proven for smooth g. ∎

1.3.16 Lemma. If ν is as in Lemma 1, then $\inf \int g^p d\nu = \inf \int g^p \bar{\nu} dt$, the inf being taken over $G \cap C_0^\infty(a,b)$.

Proof. Pick s a support of the singular part of ν with $|s| = 0$ and O_n open such that $s \subseteq O_n$, $a,b \in O_n$, and $|O_n| \to 0$. Since O_n is a collection of pairwise disjoint open intervals, it is easy to construct $C_0^\infty(a,b)$ functions $\phi_{n,i}$ (each $\phi_{n,i} = 1$ off of a finite number of the intervals such that $\phi_{n,i} \to \chi_{(a,b)-O_n}$ everywhere on (a,b) with $0 \le \phi_{n,i} \le 1$, and $\phi_{n,i} = 1$ on $(a,b) - O_n$. It then follows for $g \in G \cap C_0^\infty(a,b)$ that

$$\lim_{i\to\infty} \int (g\phi_{n,i})^p d\nu = \int g^p \chi_{(a,b)-O_n} d\nu = \int_{(a,b)-O_n} g^p \bar{\nu} dt$$

and

$$\lim_{i\to\infty} \int_a^b g\phi_{n,i} \sigma dt = \int_{(a,b)-O_n} g\sigma dt ,$$

so

$$\lim_{n\to\infty} \lim_{i\to\infty} \int (g\phi_{n,i})^p d\nu = \int_a^b g^p \bar{\nu} dt$$

and

$$\lim_{n\to\infty} \lim_{i\to\infty} \int g\phi_{n,i} dt = \int_a^b g\, dt = 1 .$$

From this extract a sequence $g_k = \dfrac{g\phi_{n_k,i_k}}{\int_a^b g\phi_{n_k,i_k} \sigma dt}$, so that $g_k \in G \cap C_0^\infty(a,b)$ and

$$\lim_{k\to\infty} \int g_k^p d\nu = \int g^p \bar{\nu} dt ,$$

and the result is proven. ∎

CHAPTER 2

The results of Chapter 2 form the foundation on which Chapter 3 is built. For the most part they involve weighted analogues of important basic tools used in the study of partial differential equations.

In Section 2.1.0 the weights for several Sobolev inequalities are characterized in a very general setting. Section 2.2.0 develops the theory of weighted Sobolev spaces, weighted capacity, and weighted Sobolev inequalities in a setting appropriate for the application to differential equations. An example is developed in which Sobolev inequalities are proven having weights of the form $\operatorname{dist}^{\sigma}(x,K)$ for a class of sets K including unions of manifolds of co-dimension ≥ 2. In Sections 2.3.0 a result on "reverse Holder" inequalities is developed which implies higher integrability for functions satisfying a maximal function inequality.

2.1.0 Weighted Sobolev Inequalities

Conditions equivalent to two types of Sobolev inequalities are developed involving the dominance of measure by "capacity". It should be noted that V. G. Mazya [Ma2] has proved 2.1.7 for ν = lebesgue measure and $M = \mathbf{R}^d$ and D. R. Adams [A1-3] has done the same for higher order inequalities (as well as two-weighted inequalities for potentials). He has also shown that 2.1.9, in the special case described above, is needed only for K which are balls. After having discussed my results on Sobolev inequalities with me, Adams found an alternative proof for 2.1.7 and some cases of 2.1.20 using strong type capacitary estimates, the study of which was initiated by V. G. Mazya [MA1].

Let (M,F,ω) and (M,G,ν) be measure spaces with ω,ν positive. Let $W_0^{1,p}(\omega,\nu,M)$ be a set of real-valued F measurable functions satisfying the following properties.

2.1.1. $W_0^{1,p}(\omega,\nu,M)$ is closed under composition with $f \in N = \{f \in C^\infty(\mathbf{R}) : f(0) = 0,\ f'$ is bounded and of one sign$\}$.

There is a map $|D|$ such that $|D| : W_0^{1,p}(\omega,\nu,M) \to L^p(\nu,M)$ and

$$\begin{cases} d^{-1}\left(\int |Df \circ \phi|^p d\nu\right)^{1/p} \leq \left(\int |f'(\phi)|^p |D\phi|^p d\nu\right)^{1/p} \\ \qquad\qquad \leq d\left(\int |Df \circ \phi|^p d\nu\right)^{1/p} \end{cases} \tag{2.1.2}$$

for some fixed $d > 0$, where the notation $|D\phi| = |D|\phi$ is used.

The symbol $|D|$ is only meant to suggest the absolute value of the gradient on the classical $W^{1,p}$ space. It should be noted that $|D|$ need not be sublinear. Special cases of $W_0^{1,p}(\omega,\nu,M)$ are developed in Section 2.2.0.

Given $\phi \in W_0^{1,p}(\omega,\nu,M)$, let μ_ϕ be the finite measure defined by $d\mu_\phi = |D\phi|^p d\nu$. μ_ϕ^* will be the distribution measure of ϕ with respect to μ_ϕ, that is, $\mu_\phi^*(E) = \mu(\phi^{-1}(E))$ so that $\int_{-\infty}^{\infty} g(t)d\mu_\phi^*(t)$

$= \int g(\phi)d\mu_\phi$ for all Borel measurable g. Also let $\bar{\mu}^*_\phi$ be the density of the absolutely continuous part of μ^*_ϕ. p' will always represent the exponent conjugate to p, that is $\frac{1}{p} + \frac{1}{p'} = 1$. The proofs of the following theorems will be deferred till later.

2.1.3 Theorem. If $1 \le p \le q < \infty$ and $\phi \in W_0^{1,p}(\omega,\nu,M)$, then

$$(\int |u|^q d\omega)^{1/q} \le c(\int |Du|^p d\nu)^{1/p} \tag{2.1.4}$$

for some $c > 0$ and all $u = f \circ \phi$, $f \in N$, iff

$$\sup_{r \neq 0} \omega^{1/q}(\frac{\phi}{r} \ge 1) \left|\int_0^r \bar{\mu}^*_\phi(t)^{-1/(p-1)} dt\right|^{1/p'} = b < \infty . \tag{2.1.5}$$

The convention $\int_s^t g = -\int_t^s g$ is used for $t < s$. If c is chosen as small as possible, then $d^{-1}b \le c \le dp^{1/q}p'^{1/p'}b$.

2.1.6 Remark. Under fairly general circumstances the co-area formula [F2] can be used to give an explicit expression for $\bar{\mu}^*_\phi$. The following is a very special case.

Suppose $M = \Omega \subseteq \mathbf{R}^n$, Ω open, $|D\phi| = |\nabla\phi|$, ν is absolutely continuous with density $\bar{\nu}$, an integrable Borel function and $\phi \in C^1(\Omega)$.

$$\int_\Omega \psi|\nabla\phi| dH^n = \int_{-\infty}^{\infty} \int_{\partial\{\phi<t\}} \psi dH^{n-1} dt$$

for all positive Borel functions ψ supported in Ω so letting $\psi = |\nabla\phi|^{p-1}\bar{\nu}\chi_{\phi^{-1}(E)}$, it follows that $\bar{\mu}^*_\phi(t) = \int_{\partial\{\phi<t\}} |\nabla\phi|^{p-1}\bar{\nu} dH^{n-1}$ a.e. This can be generalized to allow M to be a "manifold" in a weak measure-theoretic sense and ϕ to be "Sobolev".

If $\mathbf{H} \subseteq W_0^{1,p}(\omega,\nu,M)$ and $\mathbf{H}$ is closed under composition with $f \in N$, then for $A \subseteq M$ let $C_{\mathbf{H},p}(A) = \inf\{\int |D\phi|^p d\nu : \phi \in \mathbf{H}, \phi \ge 1 \text{ on } A\}$.

2.1.7 Theorem. If $1 \le p \le q < \infty$, then

$$(\int |u|^q d\omega)^{1/q} \le c(\int |Du|^p d\nu)^{1/p} \tag{2.1.8}$$

for some $c > 0$ and all $u \in \mathbf{H}$ iff

$$\omega^{1/q}(A) = bC^{1/p}_{\mathbf{H},p}(A) \tag{2.1.9}$$

for some $b > 0$, and for all sets $A = \{\frac{\phi}{r} \ge 1\}$, $\phi \in \mathbf{H}$, $r \neq 0$.

If both c,b are chosen as small as possible, then $d^{-3}b \le c \le d^3p^{1/q}p'^{1/p'}b$.

2.1.10 Remark. Although it seems in most cases that the use of the capacity $C_{\mathbf{H},p}$ is more practical, there are specific cases when the equivalent expression $K_{\mathbf{H},p}$ described in Section 1.3.0 is more easily calculated. This occurs, for instance, when the level sets of the Sobolev functions considered are of a fixed geometry or if they display certain symmetries. It is necessary in these cases to use the co-area formula, as described in 2.1.6, to calculate $\bar{\mu}_{\phi}^{*}$.

Let $(M,\mathbf{F},\lambda)$ be a measure space with λ positive, and let $W^{1,p}(\omega,\nu,M)$ be a set of real-valued $\mathbf{F}$ measurable functions satisfying the following.

2.1.11. $W^{1,p}(\omega,\nu,M)$ is closed under composition with $f \in N' = \{f \in C^{\infty}(\mathbf{R}) : f'$ is bounded and of one sign$\}$, and there is a map $|D| : W^{1,p}(\omega,\nu,M) \to L^{p}(\nu,M)$ such that (2.1.2) holds.

2.1.12 Theorem. If $1 < p < q < \infty$ and $\phi \in W^{1,p}(\omega,\nu,M)$, then

$$\left(\int \left(\int |u(x) - u(y)|\,d\lambda(y)\right)^{q} d\omega(x)\right)^{1/q} < c_1\left(\int |Du|^{p} d\nu\right)^{1/p} \qquad (2.1.13)$$

for some $c_1 > 0$ and all $u = f \circ \phi$, $f \in N'$ iff

$$\sup_{r,\psi=\pm\phi} \omega^{1/q}(\psi < r)\left(\int_{r}^{\infty} \left(\frac{\lambda^{p}(\{\psi > t\})}{\bar{\mu}_{\phi}^{*}(t)}\right)^{1/(p-1)} dt\right)^{1/p'} = b_1 < \infty . \qquad (2.1.14)$$

If $\lambda(M) = 1$, $\omega(M) < \infty$, and $\phi \in L^{1}(\lambda,M)$, then

$$\left(\int \left|u(x) - \int u(y)\,d\lambda(y)\right|^{q} d\omega(x)\right)^{1/q} < c_2\left(\int |Du|^{p} d\nu\right)^{1/p} \qquad (2.1.15)$$

for some $c_2 > 0$ and all $u = f \circ \phi$, $f \in N^{1}$ iff (2.1.14) holds.

The conventions $0 \cdot \infty = 0$ and, for $\lambda^{p}/\bar{\mu}_{\phi}^{*}$, $\frac{0}{0} = 0$, $\frac{\infty}{\infty} = 0$ are used. If the c_i are chosen as small as possible, then $d^{-1}b_1 < c_2 < c_1 < 2dp^{1/q}p'^{1/p'}b_1$.

If $\mathbf{H} \subseteq W^{1,p}(\omega,\nu,M)$ and $\mathbf{H}$ is closed under composition with $f \in N'$, then for $A \subseteq M$ let

$$\bar{C}_{\mathbf{H},p}(A) = \inf\{\int |D\phi|^{p} d\nu : \phi \in \mathbf{H} \cap L^{1}(\lambda,M)\ ,\ \phi > 1 \text{ on } A \text{ and } \int \phi\, d\lambda = 0\} ,$$

and

$$\bar{C}'_{\mathbf{H},p}(A) = \inf\{\int |D\phi|^{p} d\nu : \phi \in \mathbf{H} \cap L^{1}(\lambda,M)\ ,\ \phi < 0 \text{ on } A \text{ and } \int \phi\, d\lambda = 1\} .$$

In Theorem 2.1.17 it will be assumed that

$$\text{if } \phi \in \mathbf{H}, \text{ then } \lambda(\{\phi > t\}) < \infty \text{ for } t > 0 . \qquad (2.1.16)$$

If this is not the case, then the theorem still holds but $\bar{C}_{H,p}$ and $\bar{C}'_{H,p}$ must be replaced by the set function $\bar{K}_{H,p}$ defined in Section 1.3.0.

2.1.17 Theorem. If $1 < p < q < \infty$ and (2.1.16) holds, then

$$\left(\int \left(\int |u(x) - u(y)| d\lambda(y)\right)^q d\omega(x)\right)^{1/q} < c_1\left(\int |Du|^p d\nu\right)^{1/p} \qquad (2.1.18)$$

for some $c_1 > 0$ and all $u \in H$ iff

$$\omega^{1/q}(A) < b_1 \bar{C}'^{1/p}_{H,p}(A) \qquad (2.1.19)$$

for some $b_1 > 0$ and all sets $A = \{\phi < 0\}$, $\phi \in H$.

If $\lambda(M) = 1$, $\omega(M) < \infty$ and $H \subseteq L^1(\lambda, M)$, then

$$\left(\int |u(x) - \int u(y) d\lambda(y)|^q d\omega(x)\right)^{1/q} < c_2\left(\int |Du|^p d\nu\right)^{1/p} \qquad (2.1.20)$$

for some $c_2 > 0$ and all $u \in H$ iff

$$\omega^{1/q}(A) < b_2 \bar{C}^{1/p}_{H,p}(A) \qquad (2.1.21)$$

for some $b_2 > 0$ and all sets $A = \{\phi < 0\}$, $\phi \in H$, iff (2.1.19) holds since $d^{-p}\bar{C}_{H,p}(A) < \bar{C}'_{H,p}(A) < d^p \bar{C}_{H,p}(A)$.

If c_i, $i = 1,2$, b_i, $i = 1,2$ are chosen as small as possible, then $d^{-3} b_i < c_i < 2d^3 p^{1/q} p'^{1/p'} b_i$, $i = 1,2$.

Remark: 2.1.10 is applicable to $\bar{K}_{H,p}$, $\bar{C}'_{H,p}$, and $\bar{C}_{H,p}$ as well as $K_{H,p}$ and $C_{H,p}$.

Theorem 2.1.22 is an example of how the conditions in Theorems 2.1.7 and 2.1.17 can be put into a more computable form when $p = 1$.

2.1.22 Theorem. Let $H = C_0^\infty(\Omega)$, where $\Omega \subseteq \mathbf{R}^n$ is open, and let ν be absolutely continuous with density $\bar{\nu} \in L^1(\Omega)$.

If $p = 1$, then condition (2.1.9) is equivalent to

$$\omega^{1/q}(A) < c \liminf_{\delta \to 0} \frac{1}{\delta} \int_{C_\delta} \bar{\nu} dx \qquad (2.1.23)$$

for some $c > 0$, all A compact with C^∞ boundary and $C_\delta = \{x \notin A: \operatorname{dist}(x,A) < \delta\}$.

If $\bar{\nu}$ is continuous, then this reduces to

$$\omega^{1/q}(A) < c \int_{\partial A} \bar{\nu} dH^{n-1}, \qquad (2.1.24)$$

or, in a more suggestive notation,

$$\omega_n^{1/q}(A) < c\nu_{n-1}(\partial A) .$$

If $H = C^{\infty}(\Omega')|_{\Omega}$, Ω' open and $\bar{\Omega} \subseteq \Omega'$, then for $p = 1$ condition (2.1.19) is equivalent to

$$\omega^{1/q}(A)\lambda(\Omega - A) \leq c \liminf_{\delta \to 0} \frac{1}{\delta} \int_{C_\delta} \bar{\nu} dx \tag{2.1.25}$$

for some $c > 0$ and all A, closed relative to Ω, which extend to compact sets with C^{∞} boundary in Ω'.

If $\bar{\nu}$ is continuous, then this becomes

$$\omega^{1/q}(A)\lambda(\Omega - A) \leq c \int_{\partial A \cap \Omega} \bar{\nu} dH^{n-1} . \tag{2.1.26}$$

The proof of Theorem 2.1.22 will rely on the following proposition.

<u>2.1.27 Proposition</u>. If $\phi \in C_0^2(\Omega)$, $\Omega \subseteq R^n$ is open, $t \in R$ is such that $\{\phi = t\} \cap \{\nabla\phi = 0\} = \emptyset$, and if ω is continuous, then

$$\lim_{\delta \to 0} \frac{1}{\delta} \int_{C_\delta(t)} \omega dx = \int_{\{\phi = t\}} \omega dH^{n-1} ,$$

where $C_\delta(t) = \{x \in \{\phi < t\} : \text{dist}(x, \{\phi = t\}) < \delta\}$.

If $\phi \in C_0^n(\Omega)$ and ω is an integrable Borel measurable function, then

$$\liminf_{\delta \to 0} \frac{1}{\delta} \int_{C_\delta(t)} \omega dx \leq \int_{\{\phi = t\}} \omega dH^{n-1}$$

for almost all $t \in R$.

<u>Proof of Theorem 2.1.3</u>. Assume (2.1.4) holds. Let $\omega_\phi^*(E) = \omega(\phi^{-1}(E))$ for $E \subseteq R$ so that for all $f \in N$ and $u = f \circ \phi$,

$$\left(\int |u|^q d\omega\right)^{1/q} = \left(\int_{-\infty}^{\infty} |f(t)|^q d\omega_\phi^*(t)\right)^{1/q} \tag{2.1.28}$$

$$= \left(\int_{-\infty}^{\infty} \left|\int_0^t f'(s) ds\right|^q d\omega_\phi^*(t)\right)^{1/q}$$

since $f(0) = 0$ and the convention $\int_s^t = -\int_t^s$ is used. Also,

$$\left(\int |Du|^p d\nu\right)^{1/p} \leq d\left(\int |f'(\phi)|^p |D\phi|^p d\nu\right)^{1/p}$$
$$= d\left(\int_{-\infty}^{\infty} |f'(t)|^p d\mu_\phi^*\right)^{1/p} . \tag{2.1.29}$$

Letting $g = |f'|$ and recalling that f' does not change sign, it follows that

$$\left(\int_{-\infty}^{\infty} \left|\int_0^t g(s)ds\right|^q d\omega_\phi^*\right)^{1/q} \leq cd\left(\int_{-\infty}^{\infty} g^p(t)d\mu_\phi^*(t)\right)^{1/p} \tag{2.1.30}$$

for all bounded nonnegative C^∞ functions g.

μ_ϕ^* is a finite measure since $|D\phi| \in L_\nu^p$. Also t^q is seen to be ω_ϕ^* integrable by letting $g = 1$ in (2.1.30). Taking uniformly bounded pointwise limits of bounded nonnegative C^∞ functions g it follows that (2.1.30) holds for all bounded nonnegative Borel measurable g. Taking monotone limits then gives (2.1.30) for all nonnegative Borel measurable functions g. Using ~~(2.1.14)~~ (1.2.14) it follows that

$$\begin{cases} \sup_{0<r} \omega_\phi^{*1/q}([r,\infty))\left(\int_0^r \bar{\mu}_\phi^*(t)^{-1/(p-1)}dt\right)^{1/p'} \leq cd \\ \text{and} \\ \sup_{r<0} \omega_\phi^{*1/q}((-\infty,r])\left(\int_r^0 \bar{\mu}_\phi^*(t)^{-1/(p-1)}dt\right)^{1/p'} \leq cd . \end{cases} \tag{2.1.31}$$

But $\omega_\phi^*([r,\infty)) = \omega(\{\phi \geq r\})$ and $\omega_\phi^*((-\infty,r]) = \omega(\{\phi \leq r\})$, so considering the sign of r and using the convention $\int_s^t = -\int_t^s$, it follows that

$$\sup_{r\neq 0} \omega^{1/q}\left(\frac{\phi}{r} \geq 1\right)\left|\int_0^r \bar{\mu}_\phi^*(t)^{-1/(p-1)}dt\right|^{1/p'} \leq cd .$$

Assume (2.1.5). As above, (2.1.5) is equivalent to (2.1.31) so, by (1.2.15), (2.1.30) holds (with a different constant) for all nonnegative Borel measurable g. Given $f \in N$, let $g = |f'|$ and use (2.1.28), part of (2.1.29), and (2.1.2) to get (2.1.4). ∎

<u>Proof of Theorem 2.1.7.</u> By Theorem 2.1.3 it follows that $d^{-1}b' \leq c \leq dp^{1/q}p'^{1/p'}b'$ if c is the smallest constant in (2.1.8) and

$$b' = \sup_{\phi \in H} \sup_{r\neq 0} \omega^{1/q}\left(\frac{\phi}{r} \geq 1\right)\left|\int_0^r \bar{\mu}_\phi^*(t)^{-1/(p-1)}dt\right|^{1/p'} .$$

Given $\phi \in H$ and $r \in \mathbf{R}$, $r \neq 0$ let $f(t) = \frac{t}{r}$ so $f \in N$. $a \sim_c b$ will be used to mean that $ac^{-1} \leq b \leq ac$. $\mu_{f\circ\phi}^*(E) = \mu_{f\circ\phi}(\{f \circ \phi \in E\}) = \int_{\phi \in rE} |Df \circ \phi|^p d\nu \sim_{d^p} |r|^{-p} \int_{\phi\in rE} |D\phi|^p d\nu = |r|^{-p}\mu_\phi^*(rE)$, so $\int_E \bar{\mu}_{f\circ\phi}^*(t)dt \sim_{d^p} |r|^{-p} \int_{rE} \bar{\mu}_\phi^*(s)ds$. Divide by $|E|$ and differentiate using Lebesgue's theorem to get that $\bar{\mu}_{f\circ\phi}^*(t) \sim_{d^p} |r|^{1-p}\bar{\mu}_\phi^*(rt)$ a.e. A change of variable now gives

$$\omega^{1/q}(\{\frac{\phi}{r} > 1\}) \,|\int_0^r \bar{\mu}_\phi^*(t)^{-1/(p-1)}dt|^{1/p'}$$

$$\sim_d \omega^{1/q}(f(\phi) > 1)(\int_0^1 \bar{\mu}_{f(\phi)}^*(t)^{-1/(p-1)}dt)^{1/p'}$$

so that

$$b \sim_d \sup_{\phi\in H} \omega^{1/q}(\{\phi > 1\})(\int_0^1 \bar{\mu}_\phi^*(t)^{-1/(p-1)}dt)^{1/p'}$$

since $f \in N$. Using the $0\cdot\infty = 0$ convention, it is clear that $b' \sim_d b"$ if $b"$ is the smallest possible constant in the inequality

$$\omega^{1/q}(A) \le b" \inf\{(\int_0^1 \bar{\mu}_\phi^*(t)^{-1/(p-1)}dt)^{-1/p'} : \phi \ge 1 \text{ on } A \text{ and } \phi \in H\}$$

considered for all sets $A = \{\psi \ge 1\}$ where ψ is a function in H. Using Theorem 1.3.2 it then follows that (2.1.8) and (2.1.9) are equivalent and $d^{-3}b \le c \le d^3p^{1/q}p'^{1/p'}b$ for b, c chosen as small as possible. ∎

<u>Proof of Theorem 2.1.12</u>. As in Theorem 2.1.3, (2.1.13) reduces to

$$(\int_{-\infty}^\infty (\int_{-\infty}^\infty |\int_s^t g(\sigma)d\sigma|d\lambda_\phi^*(s))^q d\omega_\phi^*(t))^{1/q} \tag{2.1.32}$$

$$\le c_1 d(\int_{-\infty}^\infty g^p(t)d\mu_\phi^*(t))^{1/p}$$

with $g = |f'|$ and λ_ϕ^* defined as $\lambda_\phi^*(E) = \lambda(\phi^{-1}(E))$ for $E \subseteq R$.

Also, as in the proof of Theorem 2.1.3, the function $|t - s|$ has the necessary integrability properties to allow the taking of limits, thus giving (2.1.32) for all nonnegative Borel measurable g. Using (1.2.23) and arguing as in Theorem 2.1.3 it is seen that (2.1.13) is equivalent to

$$\sup_r \omega^{1/q}(\{\phi \le r\})(\int_r^\infty (\frac{\lambda^p(\{\phi \ge t\})}{\bar{\mu}_\phi^*(t)})^{1/(p-1)}dt)^{1/p'} < \infty$$

combined with

$$\sup_r \omega^{1/q}(\{\phi \ge r\})(\int_{-\infty}^r (\frac{\lambda^p(\phi \le t)}{\bar{\mu}_\phi^*(t)})^{1/(p-1)}dt)^{1/p'} < \infty .$$

It is easy to see that this is just (2.1.14).

If $\lambda(M) = 1$, $\omega(M) < \infty$ and $\phi \in L^1(\lambda, M)$, then (2.1.15) reduces to

$$\left(\int_{-\infty}^{\infty} \left|\int_{-\infty}^{\infty} \left(\int_s^t g\,d\sigma\; d\lambda_\phi^*(s)\right|^q d\omega_\phi^*(t)\right)^{1/q} \tag{2.1.33}$$

$$\leq c_2 d\left(\int_{-\infty}^{\infty} g^p(t)\,d\mu_\phi^*(t)\right)^{1/p} .$$

The finiteness of λ, ω implies that of λ_ϕ^* and ω_ϕ^* so that taking limits of C^∞ functions of compact support it is seen that (2.1.33) holds for all bounded Borel measurable g of compact support. Using (1.2.24) and continuing as above, the equivalence of (2.1.14), (2.1.15) is proven. ∎

<u>Proof of Theorem 2.1.17.</u> By Theorem 2.1.12, it follows that $d^{-1}b \leq c_1 \leq 2dp^{1/q}p'^{1/p'}b$ if c_1 is the smallest constant in (2.1.18) and

$$b = \sup_{\phi \in \mathbf{H}} \sup_r \omega^{1/q}(\{\phi \leq r\})\left(\int_0^r \left(\frac{\lambda^p(\{\phi > t\})}{\bar{\mu}_\phi^*(t)}\right)^{1/(p-1)} dt\right)^{-1/p'}$$

since $f \in N'$ if $f(t) = -t$.

Given $\phi \in \mathbf{H}$, $r \in \mathbf{R}$, let $f(t) = t - r$ so $f \in N'$ and $\mu_\phi^*(E + r) = \mu_\phi(\{\phi \in E + r\}) = \mu_\phi(\{\phi - r \in E\}) \sim_{d^p} \mu_{\phi-r}^*(E)$, and therefore, by differentiation,

$$\bar{\mu}_\phi^*(t + r) \sim_{d^p} \bar{\mu}_{\phi-r}^*(t) = \bar{\mu}_{f(\phi)}^*(t) \quad \text{a.e.}$$

(Recall μ_ϕ^* is a finite measure so $\bar{\mu}_\phi^*$ is Lebesgue integrable.) A change of variables now gives

$$\omega^{1/q}(\phi \leq r)\left(\int_r^{\infty} \left(\frac{\lambda^p(\{\phi > t\})}{\bar{\mu}_\phi^*(t)}\right)^{1/(p-1)} dt\right)^{1/p'}$$

$$\sim_d \omega^{1/q}(\{f(\phi) \leq 0\})\left(\int_0^{\infty} \left(\frac{\lambda^p(\{f(\phi) > t\})}{\bar{\mu}_{f(\phi)}^*(t)}\right)^{1/(p-1)} dt\right)^{1/p'}$$

and so $\sup_{\phi \in \mathbf{H}} \omega^{1/q}(\{\phi \leq 0\})\left(\int_0^{\infty} \left(\frac{\lambda^p(\{\phi > t\})}{\bar{\mu}^*(t)}\right)^{1/(p-1)} dt\right)^{1/p'} \sim_d b.$

Using the $0\cdot\infty = 0$ convention, it is clear that $b \sim_d b'$ if b' is the smallest constant possible in the inequality

$$\omega^{1/q}(A) \leq b' \inf\left\{\left(\int_0^{\infty} \left(\frac{\lambda^p(\{\phi > t\})}{\bar{\mu}^*(t)}\right)^{1/(p-1)} dt\right)^{-1/p'} : \phi \leq 0 \text{ on } A,\ \phi \in \mathbf{H}\right\}$$

considered for all sets $A = \{\psi \leq 0\}$ where ψ is a function in $\mathbf{H}$.

Noting (2.1.16) and using Theorem 1.3.5, it now follows that (2.1.18) and (2.1.19) are equivalent and $d^{-3}b_1 \le c_1 \le 2d^3 p^{1/q} p'^{1/p'} b_1$ if b_1 and c_1 are chosen as small as possible. The equivalence of (2.1.20) and (2.1.21) follows in a virtually identical manner. ▪

Proof of Proposition 2.1.17. If $\{\phi = t\} \cap \{|\nabla\phi| = 0\} = \emptyset$, then $|\nabla\phi| \ge \delta > 0$ on $\{\phi = t\}$ for some $\delta > 0$ since ϕ has compact support, so $M_t = \{\phi = t\}$ is an oriented compact $n - 1$ dimensional manifold. If $f_s(x) = x + n_x s$, where n_x is the unit normal to $\{\phi = t\}$ at x directed into $\{\phi < t\}$, then $\exists\, d > 0$ such that if $0 \le s \le d$, then $f_s : M_t \to f_s(M_t)$ is a diffeomorphism and $|J_s| \to 1$ as $s \to 0$, where j_s is the Jacobian of the transformation.

If $\delta \le d$ and ψ is continuous, then

$$\left|\frac{1}{\delta}\int_{C_\delta(t)} \psi\, dH^n - \int_{\{\phi=t\}} \psi\, dH^{n-1}\right|$$

$$= \left|\frac{1}{\delta}\int_0^\delta \int_{f_s(\{\phi=t\})} \psi(x)\, dH^{n-1} ds - \int_{\{\phi=t\}} \psi\, dH^{n-1}\right|$$

the co-area formula and the fact that the gradient of the distance function has absolute value one a.e. on $C(\delta)$

$$\le \frac{1}{\delta}\int_0^\delta \int_{\{\phi=t\}} |\psi(f_s(x))|J_s| - \psi(x)|\, dH^{n-1} ds$$

$$\le \varepsilon \int_{\{\phi=t\}} dH^{n-1}$$ if $\delta \le \delta_\varepsilon$ for some δ_ε since $|J_s| \to 1$ and ψ is continuous.

Therefore the first statement of the proposition is proven with $\psi = \omega$.

If $\phi \in C_0^n(\Omega)$ and $B = \{t : \{\phi = t\} \cap \{|\nabla\phi| = 0\} \neq \emptyset\}$, then the Morse-Sard theorem says that $|B| = 0$. B is closed since ϕ has compact support so $R - B = \bigcup_i I_i$, I_i being pairwise disjoint open intervals. Given $[t_0, t_1] \subseteq I_i$, then $|\nabla\phi| \ge \delta > 0$ on $\phi^{-1}([t_0, t_1])$. If f_s is defined essentially the same as before, then $f_s : \phi^{-1}(t_0, t_1) \to f_s(\phi^{-1}(t_0, t_1))$ is a diffeomorphism for all sufficiently small s, say $s \le d$ for some d.

If ψ is an integrable Borel function, then

$$\left|\int_{t_0}^{t_1} \frac{1}{\delta}\int_{C_\delta(t)} \psi\, dH^{n-1} dt - \int_{t_0}^{t_1} \int_{\phi=t} \psi\, dH^{n-1} dt\right|$$

$$\leq \int_{t_0}^{t_1} \frac{1}{\delta}\int_0^{\delta} \int_{\phi(f_s^{-1}(x))=t} \psi(x)\,dH^{n-1}ds\,dt + \int \psi|\nabla\phi|\,dH^n$$

$$\leq \frac{1}{\delta}\int_0^{\delta} \int \psi|\nabla\phi \circ f_s^{-1}|\,dH^n ds + \int \psi|\nabla\phi|\,dH^n$$

$$\leq C\int \psi\,dH^n$$

with C independent of t_0, t_1 if $t_0, t_1 \in [a,b] \subseteq I_i$ for fixed a, b.

Given $\varepsilon > 0$, pick $\bar{\omega}$ continuous such that $\int |\omega - \bar{\omega}| < \varepsilon$ and pick $\delta > 0$, $\delta = \delta_\varepsilon$ as in the first part of the proof. Then,

$$\left|\int_{t_0}^{t_1} \frac{1}{\delta}\int_{C_\delta(t)} \omega\,dH^n dt - \int_{t_0}^{t_1} \int_{\{\phi=t\}} \omega\,dH^{n-1}dt\right|$$

$$\leq \left|\int_{t_0}^{t_1} \frac{1}{\delta}\int_{C_\delta(t)} \bar{\omega}\,dH^n dt - \int_{t_0}^{t_1} \int_{\{\phi=t\}} \bar{\omega}\,dH^{n-1}dt\right|$$

$$+ \left|\int_{t_0}^{t_1} \frac{1}{\delta}\int_{C_\delta(t)} (\bar{\omega} - \omega)\,dH^n dt - \int_{t_0}^{t_1} \int_{\{\phi=t\}} (\bar{\omega} - \omega)\,dH^{n-1}dt\right|$$

$$\leq \varepsilon\left(\int_{t_0}^{t_1} \int_{\{\phi=t\}} dH^{n-1}dt + C\right) \leq \varepsilon\left(\int |\nabla\phi| + C\right),$$

so
$$\int_{t_0}^{t_1} \liminf_{\delta\to 0} \frac{1}{\delta}\int_{C_\delta(t)} \omega\,dH^n \leq \lim_{\delta\to 0} \frac{1}{\delta}\int_{t_0}^{t_1} \int_{C_\delta(t)} \omega\,dH^n dt$$

$$= \int_{t_0}^{t_1} \int_{\{\phi=t\}} \omega\,dH^n dt .$$

Now divide by $t_1 - t_0$ and let $t_1 \to t_0$ to get the final result. ∎

<u>Proof of Theorem 2.1.22.</u> Assume (2.1.23), that is

$$\omega^{1/q}(K) \leq c \liminf_{\delta\to 0} \frac{1}{\delta}\int_{C_\delta} \bar{\nu}\,dx$$

for all K compact with C^∞ boundary.

Take $\bar{\nu}$ to be a representative of the L^1 equivalence class which is Borel measurable and everywhere defined. Let $A = \{\psi \geq 1\}$ for some $\psi \in C_0^\infty(\Omega)$. Given $\phi \in C_0^\infty(\Omega)$ such that $\phi \geq 1$ on A, it follows by the Morse-Sard theorem that $\{\phi = t\} \cap \{|\nabla\phi| = 0\} = \emptyset$ for almost all $t \in \mathbf{R}$ for which it then follows that $\{\phi = t\}$ is compact with C^∞ boundary, so

$$\omega^{1/q}(A) \le \omega^{1/q}(\{\phi > 1\}) \le \omega^{1/q}(\{\phi > t\}) \quad \text{for} \quad 0 < t < 1$$

$$\le c \liminf_{\delta \to 0} \frac{1}{\delta} \int_{C_\delta(t)} \bar{\nu}dx \quad \text{a.e.,}$$

where $C_\delta(t) = \{x \in \{\phi < t\} : \operatorname{dist}(x, \{\phi = t\}) < \delta\}$, by taking $K = \{\phi = t\}$ for those t in $(0,1)$ where $\{\phi = t\} \cap \{|\nabla\phi| = 0\} = \emptyset$.

It now follows that

$$\omega^{1/q}(A) \le \inf_{(0,1)} \liminf_{\delta \to 0} \frac{1}{\delta} \int_{C_\delta(t)} \bar{\nu}dx$$

$$\le \inf_{(0,1)} \int_{\{\phi=t\}} \bar{\nu}dx \quad \text{by Proposition 2.1.27}$$

$$= \inf_{(0,1)} \bar{\mu}^*_\phi$$

considering the remark given after Theorem 2.1.3. Using Theorem 1.3.2 now shows that (2.1.9) holds for $p = 1$.

Assume (2.1.9) so that

$$\left(\int \phi^q d\omega\right)^{1/q} \le c \int |\nabla\phi| \bar{\nu}dx \tag{2.1.34}$$

for all $\phi \in C_0^\infty(\Omega)$. Given A compact with C^∞ boundary, $\exists\, \varepsilon^* > 0$ such that $\psi(x) = \operatorname{dist}(x,A)$ is C^∞ for $x \in \{0 < \operatorname{dist}(x,A) < \varepsilon^*\}$ with $|\nabla\psi(x)| = 1$. Let

$$f_\delta(x) = \begin{cases} 1 & \text{on } A\,, \\ \left(1 - \frac{\psi(x)}{\delta}\right) & \text{if } \psi < \delta, \quad x \notin A\,, \\ 0 & \text{otherwise}\,, \end{cases}$$

and let $\Phi_{n,\delta} = h_n * f_\delta$, where $h_n(x) = 2nh(2n)$, $h \in C_0^\infty(\mathbf{R})$, $h \ge 0$, $\int h = 1$, and the support of $h = \operatorname{spt} h \subseteq [-1,1]$. If $\sigma \in C_0^\infty(\Omega)$ with $\sigma \ge 1$ on a neighborhood of A, n is large and δ small, then σ dominates the $\Phi_{n,\delta}$. (2.1.34) then implies that $\sigma \in L^1(\omega,\Omega)$ since $\bar{\nu} \in L^1(\Omega)$, so the dominated convergence theorem can be used on (2.1.34) to show that $\left(\int f_\delta^q d\omega\right)^{1/q} \le c \frac{1}{\delta} \int_{C_\delta} \bar{\nu}dx$. Taking the $\liminf_{\delta \to 0}$ gives (2.1.23).

To prove (2.1.25) equivalent to (2.1.19), first extend ν, ω, λ to be zero in $\Omega' - \Omega$ and do all further work in Ω'. Assume (2.1.25) so that

$$\omega^{1/q}(A)\lambda(\Omega' - A) \le c \liminf_{\delta\to 0} \frac{1}{\delta} \int_{C_\delta} \bar{\nu}dx \tag{2.1.35}$$

for all $A \subseteq \Omega'$, A compact with C^∞ boundary.

Let $K = \{x \in \Omega : \psi(x) \le 0\}$ for some $\psi \in C^\infty(\Omega')|_\Omega$. Given $\phi \in C_0^\infty(\Omega')|_\Omega$ such that $\phi \le 0$ on K, let $\phi_1 = \sigma\phi_2$, where ϕ_2 is an extension of ϕ to Ω' and $\sigma \in C_0^\infty(\Omega')$, $\sigma = 1$ on Ω. By the Morse-Sard theorem $\{\phi_1 = t\} \cap \{|\nabla\phi_1| = 0\} = \emptyset$ for almost all t, and so $\{\phi_1 \le t\}$ is a compact set with C^∞ boundary for almost all t, but then

$$\omega^{1/q}(K) \le \omega^{1/q}(\{\phi_1 \le 0\}) \le \omega^{1/q}(\phi_1 \le t) \quad \text{for} \quad 0 \le t < \infty$$

$$\le \frac{\liminf_{\delta\to 0} \frac{1}{\delta} \int_{C_\delta(t)} \bar{\nu}dx}{\lambda(\{\phi_1 > t\})}$$ a.e. by (2.1.35), where the convention $\frac{0}{0} = \frac{\infty}{\infty} = \infty$ is used for this ratio,

$$= \frac{\liminf_{\delta\to 0} \frac{1}{\delta} \int_{C_\delta(t)} \bar{\nu}dx}{\lambda(\{\phi_1 \ge t\})}$$ a.e.,

since $\lambda(\{\phi_1 > t\})$, being monotone, has at most a countable number of discontinuities and therefore $\lambda(\{\phi_1 = t\}) = 0$ a.e.

It now follows that

$$\omega^{1/q}(K) \le \inf_{(0,\infty)} \frac{\liminf_{\delta\to 0} \frac{1}{\delta} \int_{C_\delta(t)} \bar{\nu}dx}{\lambda(\{\phi_1 \ge t\})}$$

$$\le \inf_{(0,\infty)} \frac{\int_{\{\phi_1 = t\}} \bar{\nu}dH^{n-1}}{\lambda(\{\phi_1 \ge t\})} \quad \text{by Proposition 2.1.17}$$

$$= \inf_{(0,\infty)} \frac{\mu_\phi^*(t)}{\lambda(\{\phi \ge t\})}$$

since $\bar{\nu}$ and λ are zero in $\Omega' - \Omega$ and $\phi_1|_\Omega = \phi$. Using Theorem 1.3.5 it can be seen that (2.1.19) is verified.

Assume (2.1.19), so for $\phi \in C^\infty(\Omega')|_\Omega$ it follows that

$$\left(\int \left(\int |\phi(x) - \phi(y)| d\lambda(y)\right)^q d\omega(y)\right)^{1/q} \le c \int |\nabla\phi|\bar{\nu}dx . \tag{2.1.36}$$

Given $A \subseteq \Omega'$ with C^∞ boundary, let $\phi_{n,\delta}$ be as before, recalling that the diameter of $\operatorname{spt} h_n = \frac{1}{n}$. Let

$F_n = \{x \in \Omega' : \text{dist}(x,\partial A) < \delta + \frac{1}{n}\}$. Then $\phi_{n,\delta} = 0$ on $\Omega' - A - F_n$ and $\phi_{n,\delta} \geq 1 - \frac{1}{n\delta}$ on A for $\frac{1}{n} < \delta$ since $f_\delta \geq 1 - \frac{1}{n\delta}$ on $\{x \in \Omega' : \text{dist}(x,\partial A) < 1/n\}$, and so $\delta_n * f_\delta \geq \left(1 - \frac{1}{n\delta}\right) \int h_n = 1 - \frac{1}{n\delta}$. From (2.1.36) it follows that

$$\omega^{1/q}(A)\lambda(\Omega' - A - F_n)\left(1 - \frac{1}{n\delta}\right)$$

$$\leq \left(\int_A \left(\int_{\Omega'-A-F_n} |\phi_{n,\delta}(x) - \phi_{n,\delta}(y)|\,d\lambda(y)\right)^q d\omega(x)\right)^{1/q}$$

$$\leq c \int |\nabla\phi_{n,\delta}|\bar{\nu}dx\ .$$

Let $n \to \infty$ to get

$$\omega^{1/q}(A)\lambda(\Omega' - A) \leq c\,\frac{1}{\delta}\int_{C_\delta} \bar{\nu}dx\ ,$$

and taking the $\liminf_{\delta\to 0}$ gives (2.1.25), as required.

In case $\bar{\nu}$ is continuous (2.1.24) and (2.1.26) can be shown equivalent to (2.1.9) and (2.1.19), respectively, by going through the proof above, using the first part of Proposition 2.1.27 and replacing $\liminf_{\delta\to 0}$ by $\lim_{\delta\to 0}$. Alternately, (2.1.24) and (2.1.26) may be shown equivalent by using directly the methods of Proposition 1.

2.2.0 Properties of Sobolev Spaces, Capacities and Sobolev Inequalities for Applications to Differential Equations

The Sobolev spaces and "capacities" dealt with in Sections 1.3.0 and 2.1.0 will now be placed in a setting appropriate for the applications to differential equations developed in Chapter 3.

Basic properties of the capacity $C_{\mathbf{H,P}}$ and its extremals are developed such as subadditivity and capacitability. It is shown that Sobolev spaces are closed under operations such as composition with certain Lipschitz functions. The weight conditions for Sobolev inequalities developed in Section 2.1.0 are translated into the setting of Euclidean space and an example is given, where it is shown that weights of the form $\text{dist}^\sigma(x,K)$ are admissible, for a class of sets K including unions of C^2 compact manifolds of codimension ≥ 2. The notion of quasicontinuity is developed and applied to prove a weighted analogue of a result of Bagby [BG] which characterizes $W_0^{1,p}(\Omega)$. This in turn is used to demonstrate the equivalence of two approaches to the definition of weak boundary values for the Dirichlet problem. Many of these results are true in a more general setting.

Throughout Section 2.2.0 Ω will be an open subset of R^d, $p > 1$, and ω, ν, and λ will be locally finite positive Borel measures on Ω with ν absolutely continuous to ω and $\lambda(\Omega) = 1$.

2.2.1 Sobolev Spaces. Let $\mathbf{L}^p(E) = L^p(\omega,E) \times \prod_{i=1}^{d} L^p(\nu,E)$ for $E \subseteq \Omega$, E Borel measurable. Assign $\mathbf{L}^p(E)$ the norm

$$\|(f,g_1,\dots,g_d)\|_{p;E} = \left(\int_E |f|^p d\omega + \sum_{K=1}^{d} \int_E |g_K|^p d\nu\right)^{1/p} .$$

Let $L^p_{loc}(\Omega) = \{(f,g_1,\dots,g_d) : (f,g_1,\dots,g_d)|_K \in \mathbf{L}^p(K)$ for all $K \subseteq \Omega$, K compact$\}$. $L^p_{loc}(\Omega)$ is given the topology induced by the seminorms $\|\ \|_{p;K}$, $K \subseteq \Omega$ compact. $W^{1,p}(\omega,\nu,\Omega)$ is now defined as the closure of $H = \{(\phi,\nabla\phi) : \phi \in C^\infty(\Omega) \cap L^p(\omega,\Omega)$ and $\nabla\phi \in \prod_{K=1}^{d} L^p(\nu,\Omega)\}$ in $L^p(\Omega)$; $W^{1,p}_0(\omega,\nu,\Omega)$ as the closure of $H \cap C^\infty_0(\Omega) \times \prod_{K=1}^{\infty} C^\infty_0(\Omega)$ in $L^p(\Omega)$; and $W^{1,p}_{loc}(\Omega)$ as the closure of H in $L^p_{loc}(\Omega)$.

Given $(u,v) \in W^{1,p}_{loc}(\omega,\nu,\Omega)$, the notation $v = \nabla u$, $u \in W^{1,p}_{loc}(\omega,\nu,\Omega)$ and $\|(u,v)\|_{p;\Omega} = \|u\|_{1,p}$ will be used for convenience even though this is misleading. It is not claimed that u has a unique gradient. In fact Serapioni has observed that for some weighted Sobolev spaces, zero may have a nontrivial gradient in the sense above as well as a zero gradient. Under fairly weak conditions it can be shown that if (u_1,v), $(u_2,v) \in W^{1,p}(\omega,\omega,\Omega)$, then $u_1 = u_2$ almost everywhere. If $\omega(E) = 0$ on sets E of capacity zero, then this will follow from Proposition 7. For convenience $W^{1,p}(\Omega)$ will be used to denote $W^{1,p}(\omega,\nu,\Omega)$.

One of the basic operations needed in the theory of Sobolev functions is composition with Lipschitz functions. The following proposition shows that this is possible for a wide class of Lipschitz functions. For example, any Lipschitz function with at most a countable number of discontinuities in its derivative is acceptable. The other basic operations considered are needed in Chapter 3 to show that certain functions are allowable as test functions in the definition of weak solution.

Unless a particular space is specified, all the Sobolev functions in Proposition 2.2.2 will be assumed to lie in one fixed Sobolev space, the three possible cases being $W^{1,p}_{loc}(\Omega)$, $W^{1,p}(\Omega)$, and $W^{1,p}_0(\Omega)$. Convergence is always that appropriate to the particular space considered unless otherwise indicated. It will be assumed throughout that $(u,\nabla u)$, $(u_n,\nabla u_n)$, $(v,\nabla v)$, $(v_n,\nabla v_n)$ are Sobolev functions and that $f(0) = f_n(0) = 0$ if $W_0^{1,p}(\Omega)$ is being considered or if $\omega(\Omega) = \infty$ and $W^{1,p}(\Omega)$ is being considered.

2.2.2 Proposition. Assuming the above it follows that:

(2.2.3) If $f \in C^1(\mathbf{R})$ with f' bounded, and if $u_n \in C^\infty(\Omega)$ with $(u_n, \nabla u_n) \to (u, \nabla u)$, then

$$(f(u_{n_m}), f'(u_{n_m})\nabla u_{n_m}) \to (f(u), f'(u)\nabla u)$$

for some subsequence $\{n_m\}$.

(2.2.4) Suppose $f : \mathbf{R} \to \mathbf{R}$ is uniformly Lipschitz and $\exists f_n \in C^1(\mathbf{R})$ such that f_n' converges everywhere in a uniformly bounded pointwise manner to a Borel measurable function g, $g = f'$ a.e., and $f_n(0) \to f(0)$.

If $(u_n, \nabla u_n) \to (u, \nabla u)$, then there is a sequence n_m such that

$$(f_m(u_{n_m}), f_m'(u_{n_m})\nabla u_{n_m}) \to (f(u), g(u)\nabla u)$$

and if $u_{n_m} \to u$ pointwise everywhere on a set E, then $f_m(u_{n_m}) \to f(u)$ pointwise on E as well.

(2.2.5) Let

$$X_E = \begin{cases} 1 & x \in E \\ 0 & \text{otherwise} \end{cases}, \qquad x^+ = \begin{cases} x & x > 0 \\ 0 & x \leq 0 \end{cases},$$

$$\operatorname{sign} x = \begin{cases} 1 & x > 0 \\ 0 & x = 0 \\ -1 & x < 0 \end{cases}, \qquad h_{a,b}(x) = \begin{cases} b & x \geq b \\ x & a \leq x \leq b \\ a & x \leq a \end{cases},$$

where it is assumed that $a \leq 0 \leq b$ in the $W_0^{1,p}(\Omega)$ case or in the $W^{1,p}(\Omega)$ case if $\omega(\Omega) = \infty$. The cases $a = -\infty$ and $b = \infty$ are included.

For each of the pairs $(x^+, X_{\{x>0\}})$, $(|x|, \operatorname{sign} x)$, $(h_{a,b}(x), X_{\{a<x<b\}})$, represented as (f, g), there is a sequence $\{f_n\} \subseteq C^\infty(\mathbf{R})$ such that f, $\{f_n\}$, g satisfy the requirements of 2.2.4. Therefore it follows that $(u^+, X_{\{u>0\}}\nabla u)$, $(|u|, \operatorname{sign} u\nabla u)$, and $(h_{a,b}(u), X_{\{a<u<b\}}\nabla u)$ are Sobolev functions.

The f_n may be chosen to converge uniformly. For $f(x) = x^+$ or $|x|$, the f_n may be chosen such that $0 \leq f_n(x) \leq f(x)$ and for $f(x) = h_{a,b}(x)$, the f_n may be chosen such that $a \leq f_n \leq b$. If in addition $a < 0 < b$, then $a < f_n < b$ is possible.

(2.2.6) If $f : \mathbf{R} \to \mathbf{R}$, $f \in C^1[a,b]$ and $a < u < b$, then $(f(u), f'(u)\nabla u)$ is Sobolev.

(2.2.7) If u and v are bounded, then $(uv, v\nabla u + u\nabla v)$ is Sobolev.

(2.2.8) If u, v are bounded, $u \in W_0^{1,p}(\Omega)$ and $v \in W^{1,p}(\Omega)$, then $uv \in W_0^{1,p}(\Omega)$.

(2.2.9) If $u \in C_0^\infty(\Omega)$, $v \in W_{loc}^{1,p}(\Omega)$, and either v is bounded or $\nu < c\omega$, then $uv \in W_0^{1,p}(\Omega)$.

Proof of Proposition 2.2.2.

Throughout the proof it will be assumed that $E = \Omega$ if $W_0^{1,p}(\Omega)$ or $W^{1,p}(\Omega)$ are being considered and E is an arbitrary compact subset of Ω if $W_{loc}^{1,p}(\Omega)$ is being considered.

Proof of 2.2.3. Since $|f'(x)| < M$ for some $M < \infty$, then $f(x) < M|x| + f(o)$, so $f(u) \in L^p(\omega,\Omega)$ or $L_{loc}^p(\omega,\Omega)$ depending on the case being considered. Also on some subsequence n_m, $u_{n_m} \to u$ pointwise almost everywhere with respect to ω (and also ν since ν is absolutely continuous to ω), so

$$\int_E |f'(u_{n_m})\nabla u_{n_m} - f'(u)\nabla u|^p d\nu < \int_E |f'(u_{n_m})|^p |\nabla u_{n_m} - \nabla u|^p d\nu$$

$$+ \int_E |f'(u_{n_m}) - f'(u)|^p |\nabla u|^p d\nu \to 0$$

since $|f'(u_{n_m})| < M$ and $|f'(u_{n_m}) - f'(u)| \to 0$ pointwise almost everywhere ν in a uniformly bounded manner. Also

$$\int_E |f(u_{n_m}) - f(u)|^p d\omega < M^p \int_E |u_{n_m} - u|^p d\omega \to 0 \ . \ \blacksquare$$

Proof of 2.2.4. By 2.2.3, $(f_m(u_n), f_m'(u_n)\nabla u_n)$ is Sobolev, also $u_{n_i} \to u$ pointwise also everywhere ω, ν on some subsequence n_i so

$$\int_E |g(u)\nabla u - f_m'(u_{n_i})\nabla u_{n_i}|^p d\nu < \int_E |g(u) - f_m'(u)||\nabla u|^p d\nu$$

$$+ \int_E |f_m'(u)\nabla u - f_m'(u_{n_i})\nabla u_{n_i}|^p d\nu ,$$

and therefore

$$\lim_{m\to\infty} \limsup_{i\to 0} \int_E |g(u)\nabla u - f_m'(u_{n_i})\nabla u_{n_i}|^p d\nu = 0 ,$$

the second term converging to zero as in 2.2.3, and the first converging to zero since $f_m' \to g$ everywhere in a pointwise uniformly

bounded manner. A subsequence $\{\bar{n}_m\}$ can now be chosen so that
$f_m'(u_{\bar{n}_m})\nabla u_{\bar{n}_m} \to g(u)\nabla u$ in $\prod_{k=1}^{d} L^p(\nu,\Omega)$.

For $x > 0$,

$$|f_m(x) - f(x)| \le \int_0^t |f_m'(s) - g(s)|ds + |f_m(o) - f(o)|$$

$$\le M|x| + |f_m(o) - f(o)|$$

for some $M < \infty$ and also $f_m(x) \to f(x)$ since $f_m' \to g$ pointwise in a uniformly bounded manner and $f_n(o) \to f(o)$. The same is true for $x < 0$ so the dominated convergence theorem implies that $\int_E |f_m(u) - f(u)|^p d\omega \to 0$. Also $|f_m'| \le M$ for some $M < \infty$ and all m, so

$$\int_E |f_m(u_{\bar{n}_m}) - f_m(u)|^p d\omega \le M^p \int |u_{\bar{n}_m} - u|^p d\omega \to 0 .$$

Combining these shows that

$$\int_E |f_m(u_{\bar{n}_m}) - f(u)|^p d\omega \le \int_E |f_m(u) - f(u)|^p d\omega$$

$$+ \int_E |f_m(u_{\bar{n}_m}) - f_m(u)|^p d\omega \to 0$$

and so $f_m(u_{\bar{n}_m}) \to f(u)$ in Sobolev norm.

If $u_{n_m} \to u$ pointwise everywhere on a set F, then

$$|f_m(u_{n_m}) - f(u)| \le |f_m(u) - f(u)| + |f_m(u_{n_m}) - f_m(u)|$$

$$\le |f_m(u) - f(u)| + M|u_{n_m} - u| \to 0$$

on F as well. ∎

<u>Proof of 2.2.5.</u> Pick $\eta \in C_0^\infty(\mathbf{R})$ such that the support of $\eta \subseteq [0,1]$, $\eta \ge 0$, and $\int \eta = 1$. Let $\eta_n(x) = n\eta(nx)$, $f(x) = x^+$, and $f_n(x) = \eta_n * f(x)$ so that

$$f_n'(x) = \eta_n * f_n'(x) = \begin{cases} 0 & x \le 0 \\ 1 & x \ge \frac{1}{n} \end{cases}$$

with $0 \le f_n' \le 1$. It is now clear that

(2.2.10) $f_n' \to X_{\{x>0\}}$ everywhere in a pointwise uniformly bounded manner. Also $f_n(0) = f(0) = 0$ and $0 \le f_n(x) \le x^+$ since

$$0 \le \eta_n * f(x) = x^+ - \int_0^1 \eta(y)(x^+ - (x - \tfrac{y}{n})^+)dy \le x^+ .$$

The fact that the f_n converge uniformly follows from (2.2.10), $f_n(0) = f_m(0)$, and $f_n' = f_m'$ in $\mathbf{R} - [0,1]$.

For $f(x) = |x|$ use that $|x| = x^+ + (-x)^+$ in combination with the smoothing of x^+ done above to define f_n so that $f_n'(x) \to \text{sign } x$ everywhere and the conditions of b) are met.

For $f(x) = h_{a,b}(x)$ pick $\eta \in C_0^\infty(\mathbf{R})$ with $\eta \ge 0$, $\int \eta = 1$, and the support of $\eta \subseteq [-1,1]$. Let $\eta_n = n\eta(nx)$ and $g_n(x) = \eta_n * X_{E_n}$ with $E_n = \{a + \frac{2}{n} < x < b - \frac{2}{n}\}$ so

$$0 \le g_n \le 1, \; g_n(x) = \begin{cases} 0 & \text{if } \; x \le a + \frac{1}{n} \; \text{ or } \; x \ge b - \frac{1}{n} \\ 1 & \text{if } \; a + \frac{3}{n} \le x \le b - \frac{3}{n} \end{cases} \tag{2.2.11}$$

for large n and $g_n(x) \to X_{\{a<x<b\}}$ everywhere in a pointwise uniformly bounded manner.

Let

$$f_n(x) = f(0) + \int_0^x g_n(s)ds , \tag{2.2.12}$$

(where the convention $\int_0^x = -\int_x^0$ for $x < 0$ is used) so that $f_n(0) = f(0)$ and $f_n' = g_n \to X_{\{a<x<b\}}$ everywhere in a pointwise uniformly bounded manner. Uniform convergence of the f_n follows as for x^+. If $a < 0 < b$, then it is seen from(2.2.11) and (2.2.12) that $a < f_n < b$ for large n. Otherwise if follows similarly that $a \le f_n \le b$. ∎

<u>Proof of 2.2.6.</u> With f_n as above, apply 2.2.4 to $f(f_n)$, $f(h_{a,b})$, and $g = f'(h_{a,b})X_{\{a<b<b\}}$. Since $a < u < b$, this implies 2.2.6. ∎

<u>Proof of 2.2.7.</u> Assume $|u|,|v| \le M < \infty$ and apply 2.2.5 with $h_{a,b}$, $a = -M$, $b = M$ in combination with 2.2.4 to see that $\exists \, \{u_n\},\{v_n\} \in C^\infty(\Omega)$ such that $(u_n,\nabla u_n) \to (u,\nabla u)$ and $(v_n,\nabla v_n) \to (v,\nabla v)$ with $|u_n|,|v_n| \le M$. In addition, choose the sequences so that they converge pointwise almost everywhere ω,ν. Consequently,

$$\int_E |uv - u_nv_n|^p d\omega \le \int_E |u|^p|v - v_n|^p d\omega + \int_E |v_n|^p|u - u_n|^p d\omega$$

$$\le M^p(\int_E |v - v_n|^p d\omega + \int_E |u - u_n|^p d\omega) \to 0 ,$$

and

$$\left(\int_E |(u\nabla v + v\nabla u) - (u_n\nabla v_n + v_n\nabla u_n)|^p d\nu\right)^{1/p}$$

$$\leq \left(\int_E |u\nabla v - u_n\nabla v_n|^p d\nu\right)^{1/p} + \left(\int_E |v\nabla u - v_n\nabla u_n|^p d\nu\right)^{1/p}$$

$$\leq \left(\int_E |u - u_n|^p |\nabla v|^p d\nu\right)^{1/p} + M\left(\int |\nabla v - v_n|^p d\nu\right)^{1/p}$$

$$+ \left(\int_E |v - v_n|^p |\nabla u|^p d\nu\right)^{1/p} + M\left(\int |\nabla u - u_n|^p d\nu\right)^{1/p} \to 0$$

since $v_n \to v$ and $u_n \to u$ almost everywhere in a pointwise uniformly bounded manner. ∎

Proof of 2.2.8 and 2.2.9. Assume that $u \in W_0^{1,p}(\Omega)$, $v \in W^{1,p}(\Omega)$, and u, v are bounded. In the proof of 2.2.7 choose $E = \Omega$ and $u_n \in C_0^\infty(\Omega)$ so that $uv \in W_0^{1,p}(\Omega)$ since $u_n v_n \in C_0(\Omega)$.

If instead $u \in C_0^\infty(\Omega)$, $v \in W_{loc}^{1,p}(\Omega)$, and v is bounded, then choose E = support u and choose $u_n = u$, so again $uv \in W_0^{1,p}(\Omega)$. In the last case when $u \in C_0^\infty(\Omega)$, $v \in W_{loc}^{1,p}(\Omega)$ and $\nu \leq c\omega$, the only change is that $\nu \leq c\omega$ is used to show that $\int_E |v - v_n|^p |\nabla u|^p d\nu \to 0$. This is clear since $|\nabla u|$ is bounded on E = support u and $v_n \to v$ in $L^p(\omega, E)$. ∎

2.2.13 Capacity. $C_{H,p}$, $\bar{C}_{H,p}$, and $\bar{C}'_{H,p}$ will be redefined and $C_{H,p}$ will be shown to be subadditive and capacitable. The concepts of quasi-continuity and capacitary extremal will be developed. The proof of the fact that the capacitary extremal satisfies a degenerate elliptic partial differential equation will be left to later, when it is used to prove a particular Sobolev inequality.

If H, as described in 2.1.7, is $C_0^\infty(\Omega)$, then all level sets are compact and the conditions, equivalent to the Sobolev inequalities dealt with in 2.1.7, only involve capacities of compact sets. This motivates an alternate and more classical definition of capacity for noncompact sets.

Let H be a subset of $C^\infty(\Omega)$ (the functions typically vanishing on some set or a nbd of some set) closed under addition, and composition with $f \in \{f \in C^\infty : f' \text{ bounded}, \ f(0) = 0\}$. Let

$$C'_H(K) = \inf\{\int |\nabla\phi|^p d\nu : \phi \in H, \ \phi \geq 1 \text{ on } K\}$$

for $K \subset \Omega$ compact,

$$C'_H(0) = \sup\{C'_H(K) : K \subseteq 0, \ K \text{ compact}\}$$

for $O \subseteq \Omega$ open,

$$C_H(E) = \inf\{C_H'(O) : O \text{ open } E \subseteq O \subseteq \Omega\}$$

for arbitrary $E \subseteq \Omega$.

2.2.14 Proposition. C_H is monotone increasing and for E either compact or open

$$C_H'(E) = C_H(E) .$$

Proof. If E is open and $E \subseteq O$, O open, then for any $K \subseteq E$, K is also in O so $C_H'(O) \geq C_H'(E)$ and so $C_H(E) \geq C_H'(E)$, but $C_H'(E) \geq C_H(E)$ since E is open, so $C_H'(E) = C_H(E)$ as required.

If E is compact and if $E \subseteq O$, O open, then $C_H'(O) \geq C_H'(E)$, and taking the infimum over such open sets gives $C_H(E) \geq C_H'(E)$. If $\phi \in H$ and $\phi \geq 1$ on E, then $\phi > 1 - \varepsilon$ on an open set O with $E \subseteq O$, so that

$$C_H(E) \leq C_H'(\{\phi > 1 - \varepsilon\})$$

$$= \sup\{C_H'(K) : K \subseteq \{\phi > 1 - \varepsilon\},\ K \text{ compact}\}$$

$$\leq \frac{1}{(1 - \varepsilon)^p} \int |\nabla\phi| d\nu$$

since $\frac{\phi}{1 - \varepsilon} \geq 1$ on all $K \subseteq \{\phi > 1 - \varepsilon\}$. Now let $\varepsilon \to 0$ and take the infimum over all such ϕ to get $C_H(E) \leq C_H'(E)$. ∎

2.2.15 Proposition. If $A, B \subseteq \Omega$, then

$$C_H(A \cup B) + C_H(A \cap B) \leq C_H(A) + C_H(B) \tag{2.2.16}$$

and C_H is capacitable, that is, if E is Suslin (this includes the Borel sets), then $\exists K_n$ compact such that $K_n \subseteq E$ and $C_H(K_n) \to C_H(E)$ as $n \to \infty$.

Proof. Once (2.2.16) is proven, then capacitability follows from a theorem of Choquet [C]. Assume A and B are compact. If $C_H(A) = \infty$ or $C_H(B) = \infty$, then (2.2.16) holds. Otherwise let $\mathrm{SMax}_n(x,y) = f_n(y - x) + x$ and $\mathrm{SMin}_n(x,y) = y - f_n(y - x)$, where $f_n(x)$ is the smoothing of x^+ as in Proposition 2.2.2, so $f_n(x) \to x^+$, $f_n' \to \chi_{\{x>0\}}$, and $x^+ - \varepsilon_n \leq f_n(x) \leq x^+$ for some $\varepsilon_n \downarrow 0$. Pick ϕ, $\psi \in H$ with $\phi \geq 1$ on A, $\psi \geq 1$ on B, $\int |\nabla\phi|^p d\nu < \infty$ and $\int |\nabla\psi|^p d\nu < \infty$. Let $\sigma_{1,n} = \mathrm{SMax}_n(\phi,\psi)$, $\sigma_{2,n} = \mathrm{SMin}_n(\phi,\psi)$, so $\sigma_{1,n} \to (\psi - \phi)^+ + \phi = \mathrm{Max}(\phi,\psi)$,

$\sigma_{2,n} + \psi - (\psi - \phi)^+ = \mathrm{Min}(\phi,\psi)$, and $\dfrac{\sigma_{1,n}}{1 - \varepsilon_n} \geq 1$ on $A \cup B$, $\sigma_{2,n} \geq 1$ on $A \cap B$.

From Proposition 2.2.14 and the definition of C'_H it now follows that

$$C_H(A \cup B) + C_H(A \cap B)$$

$$\leq \frac{1}{(1 - \varepsilon_n)^p} \int |\nabla \sigma_{1,n}|^p d\nu + \int |\nabla \sigma_{2,n}|^p d\nu$$

$$= \frac{1}{(1 - \varepsilon_n)^p} \int |f'_n(\psi - \phi)(\nabla\psi - \nabla\phi) + \nabla\phi|^p d\nu$$

$$+ \int |\nabla\psi - f'_n(\psi - \phi)(\nabla\psi - \nabla\phi)|^p d\nu$$

$$\to \int |X_{\{\psi > \phi\}}(\nabla\psi - \nabla\phi) + \nabla\phi|^p d\nu$$

$$+ \int |\nabla\psi - X_{\{\psi > \phi\}}(\nabla\psi - \nabla\phi)|^p d\nu$$

$$= \int_{\{\psi > \phi\}} |\nabla\psi|^p d\nu + \int_{\{\psi \leq \phi\}} |\nabla\phi|^p d\nu + \int_{\{\psi > \phi\}} |\nabla\phi|^p d\nu$$

$$+ \int_{\{\psi \leq \phi\}} |\nabla\psi|^p d\nu = \int |\nabla\psi|^p d\nu + \int |\nabla\phi|^p d\nu \,.$$

Taking the infimum over such ϕ, ψ it follows that

$$C_H(A \cup B) + C_H(A \cap B) \leq C'_H(A) + C'_H(B)$$

and so (2.2.16) follows for A, B compact by Proposition 2.2.14.

If $\{K_n\}$ are compact and O is open, then $\{K_n\}$ is said to approximate O if $K_n \subseteq$ interior K_{n+1} and $\bigcup K_n = O$. Assume A, B are open and pick $\{A_n\}$, $\{B_n\}$, compact sets which approximate A, B, respectively. It is seen that $\{A_n \cup B_n\}$ and $\{A_n \cap B_n\}$ approximate $A \cup B$ and $A \cap B$, respectively. Given $K \subset A \cup B$ and $C \subseteq A \cap B$, K and C compact, then $K \subseteq A_n \cup B_n$, and $C \subseteq A_n \cap B_n$ for some n, so

$$C_H(K) + C_H(C) \leq C_H(A_n \cup B_n) + C_H(A_n \cap B_n)$$

$$\leq C_H(A_n) + C_H(B_n) \leq C_H(A) + C_H(B) \,.$$

Taking the supremum over all such K and C, and using Proposition 2.2.14, it follows that (2.2.16) holds for open sets.

Assume A, B are arbitrary sets in Ω. Given open sets O, $P \subseteq \Omega$ with $A \subseteq O$ and $B \subseteq P$, then

$$C_H(A \cup B) + C_H(A \cap B) \le C_H(O \cup P) + C_H(O \cap P)$$

$$\le C_H(O) + C_H(P)$$

and taking the infimum over such O, P shows that (2.2.16) holds. ∎

2.2.17 Proposition. C_H is countably subadditive.

Proof. Given $E_i \subseteq \Omega$, $i = 1,\ldots,n$, it follows from Proposition 2.2.16 that

$$C_H(\bigcup_{i=1}^{n} E_i) \le \sum_{i=1}^{n} C_H(E_i) .$$

Let $\{O_i\}_{i=1}^{\infty}$ be open sets and K a compact set with $K \subset \bigcup_{i=1}^{\infty} O_i$ so

$$C_H(K) \le C_H(\bigcup_{i=1}^{m} O_i) \le \sum_{i=1}^{m} C_H(O_i) \le \sum_{i=1}^{\infty} C_H(O_i)$$

for some m, so taking the supremum over all such K it follows that $C_H(\bigcup_{i=1}^{\infty} O_i) \le \sum_{i=1}^{\infty} C_H(O_i)$. Finally for $E_i \subseteq \Omega$, $i = 1,2,\ldots,$ if $C_H(E_i) = \infty$ for some i, then $C_H(\bigcup_{i=1}^{\infty} E_i) \le \sum_{i=1}^{\infty} C_H(E_i)$. Otherwise pick O_i open such that $E_i \subseteq O_i \subseteq \Omega$ and $C_H(O_i) \le C_H(E_i) + \varepsilon 2^{-i}$ so

$$C_H(\bigcup_{i=1}^{\infty} E_i) \le C_H(\bigcup_{i=1}^{\infty} O_i) \le \sum_i C_H(O_i) \le \varepsilon + \sum_i C_H(E_i)$$

and letting $\varepsilon \to 0$ gives the result. ∎

Propositions 2.2.18 and 2.2.19 will be used to motivate the definition of quasicontinuity.

2.2.18 Proposition. If $\phi \in H$, then

$$C_H(\{|\phi| > \lambda\}) \le \frac{2}{\lambda^p} \int |\nabla\phi|^p d\nu$$

for all $\lambda > 0$.

Remark. Using a smoothing of the absolute value function the coefficient 2 may be replaced by 1.

Proof. If $K \subseteq \{\phi > \lambda\}$ and K is compact, then $\frac{\phi}{\lambda} \ge 1$ on K, so $C_H(K) \le \frac{1}{\lambda^p} \int |\nabla\phi|^p d\nu$, therefore taking the supremum over all such K gives $C_H(\{\phi > \lambda\}) \le \frac{1}{\lambda^p} \int |\nabla\phi|^p d\nu$. Considering that

$\{|\phi| > \lambda\} = \{\phi > \lambda\} \cup \{-\phi > \lambda\}$ and using the subadditivity of C_H, it is seen that Proposition 2.2.18 holds. ▪

2.2.19 Proposition. If $\phi_n \in H$, $n = 1,2,\ldots,$ and the ϕ_n are Cauchy in $W^{1,p}(\Omega)$, then a subsequence of the ϕ_n converges uniformly off open sets of arbitrarily small C_H capacity.

If $H = C_0(\Omega)$, $\{\phi_n\}_{n=1}^{\infty}$ is Cauchy in $W^{1,p}_{loc}(\Omega)$ and either $|\phi_n| \le M < \infty$ or $\nu \le c\omega$, then there is a subsequence $\{n_i\}$ and there are open sets of arbitrarily small C_H capacity off of which the ϕ_{n_i} converge uniformly on compact sets. In any of these cases a subsequence of the ϕ_n converges pointwise off a set of C_H capacity zero.

Proof. Choose n_i iteratively so that $n_i < n_{i+1}$ and $\|\phi_{n_i} - \phi_m\|^p_{1,p} < 2^{-(i+2)(p+1)}$ for all $m > n_i$. Let $E_i = \{|\phi_{n_i} - \phi_{n_{i+1}}| > 2^{-(i+2)}\}$ so that by Proposition 2.2.18 it follows that $C_H(E_i) < 2^{-(i+1)}$ and so $C_H(\bigcup_{i>m} E_i) < 2^m$ for all $m > 0$. If $x \in \Omega - \bigcup_{i>m} E_i$, then $|\phi_{n_m}(x) - \phi_{n_i}(x)| \le \sum_{k=m}^{i-1} |\phi_{n_k}(x) - \phi_{n_{k+1}}(x)| < 2^{-m}$ for $i > m$ and so the ϕ_{n_i} converge uniformly off the open set $\bigcup_{i>m} E_i$ which has capacity 2^{-n}.

If $H = C_0^{\infty}(\Omega)$ and $\{\phi_n\}_{n=1}^{\infty}$ is Cauchy in $W^{1,p}_{loc}(\Omega)$, then pick $K_i \subseteq \Omega$ compact such that $K_i \subseteq$ interior K_{i+1} and $\bigcup K_i = \Omega$ and choose $\eta_i \in C_0^{\infty}$ (interior K_{i+1}) so that $\eta_i = 1$ on K_i. Also choose a subsequence n_i so that the ϕ_{n_i} converge ω almost everywhere (and so ν almost everywhere since ν is assumed absolutely continuous with respect to ω) so that

$$\int |\eta_i\phi_{n_j} - \eta_i\phi_{n_k}|^p d\omega \le c \int_{K_{i+1}} |\phi_{n_j} - \phi_{n_k}|^p d\omega \to 0$$

as $j,k \to \infty$, and

$$\int |\nabla(\eta_i\phi_{n_j}) - \nabla(\eta_i\phi_{n_k})|^p d\nu$$

$$\le \int |(\nabla\eta_i)(\phi_{n_j} - \phi_{n_k})|^p d\nu + c \int_{K_{i+1}} |\nabla\phi_{n_j} - \nabla\phi_{n_k}|^p d\nu \to 0$$

as $j,k \to \infty$ if $|\phi_{n_i}| \le M$ or if $\nu \le c\omega$. Therefore $\{\eta_i\phi_{n_j}\}_{j=1}^{\infty}$ is Cauchy in $W^{1,p}(\Omega)$ and so by the first part of Proposition 2.2.19 subsequences $n_{i,j}$ can be chosen iteratively so that $\{n_{i,j}\}$ is a subsequence of $\{n_{i-1,j}\}$, and $\{\eta_i\phi_{n_{i,j}}\}_{j=1}^{\infty}$ converges uniformly on

$K_i - G_i$, where G_i is an open set with $C_H(G_i) < 2^{-(i+1)}$. Let $F_k = \bigcup_{i>k} G_i$ so $C_H(F_k) < 2^{-k}$ and the diagonalized sequence $\phi_{n_{jj}}$ converges uniformly on any compact subset K of $\Omega - F_k$ since K must lie in K_i for some i. ∎

2.2.20 Definition. If a property holds everywhere except possibly on a set of C_H capacity zero, where $H = C_0^\infty(\Omega)$, then it is said to hold quasi-everywhere. If $u \in W_0^{1,p}(\Omega)$ and there exists a sequence $\phi_n \in C_0^\infty(\Omega)$ such that $\phi_n \to u$ in $W_0^{1,p}(\Omega)$ and $\phi_n \to u$ pointwise quasi-everywhere, then it is said that u is quasicontinuous. If $u \in W_{loc}^{1,p}(\Omega)$ and there exist $\phi_n \in C^\infty(\Omega)$ such that $\phi_n \to u$ in $W_{loc}^{1,p}(\Omega)$, then it is said that u is locally quasicontinuous.

2.2.21 Proposition

(2.2.22) If $u \in W_0^{1,p}(\Omega)$, then u can be redefined ω almost everywhere so as to be quasicontinuous.

(2.2.23) If u is quasicontinuous, then u is continuous off open sets of arbitrarily small C_H capacity for $H = C_0^\infty(\Omega)$ and if $\phi_n \in C_0^\infty(\Omega)$ and $\phi_n \to u$ in $W_0^{1,p}(\Omega)$, then $\phi_{n_i} \to u$ pointwise quasi-everywhere for some subsequence $\{n_i\}$.

(2.2.24) If either u is bounded or $\nu < c\omega$, then (2.2.22) and (2.2.23) hold for $u \in W_{loc}^{1,p}(\Omega)$ and $\phi_n \in C^\infty(\Omega)$ if quasicontinuity is replaced by local quasicontinuity.

Remark. The conditions u bounded and $\nu < c\omega$ in Propositions 2.2.19 and 2.2.21 are actually needed only near the boundary of Ω.

Proof. Given $u \in W_0^{1,p}(\Omega)$, there exist $\psi_n \in C_0^\infty(\Omega)$ such that $\psi_n \to u$ in $W_0^{1,p}(\Omega)$. Using Proposition 2.2.19 a subsequence $\{n_i\}$ may be chosen such that ψ_{n_i} converges pointwise quasi-everywhere as well as ω almost everywhere so u can be redefined on a set of ω measure zero to equal $\lim_{i\to\infty} \psi_{n_i}(x)$, where it exists, and so is quasicontinuous.

If u is quasicontinuous, then there exist $\psi_n \in C_0^\infty(\Omega)$ such that $\psi_n \to u$ in $W_0^{1,p}(\Omega)$ and pointwise off a set E, where $C_H(E) = 0$, $H = C_0^\infty(\Omega)$. By Proposition 2.2.19, there exists a subsequence $\{n_i\}$ such that ψ_{n_i} converge uniformly off open sets G_j with $C_H(G_j) < \varepsilon_j$, $\varepsilon_j \to 0$. Choose O_j open such that $E \subseteq O_j \subseteq \Omega$, and $C_H(O_j) < \varepsilon_j$ so $\psi_{n_i} \to u$ uniformly off $G_j \cup O_j$ and $C_H(G_j \cup O_j) < 2\varepsilon_j$. Therefore u is continuous off open sets of arbitrarily small C_H capacity.

If $\phi_n \in C_0^\infty(\Omega)$ and $\phi_n \to u$ in $W_0^{1,p}(\Omega)$, then $\phi_n - \psi_n \to 0$ in $W_0^{1,p}(\Omega)$ and by Proposition 2.2.19 $\phi_{n_i} - \psi_{n_i} \to 0$ quasi-everywhere for some subsequence $\{n_i\}$. $\psi_{n_i} \to u$ quasi-everywhere so it is now clear that $\phi_{n_i} \to u$ quasi-everywhere.

The proofs above go over to the $W_{loc}^{1,p}(\Omega)$ case with minor changes as in the proof of 2.2.19. ∎

It is useful to know that there exist extremal functions where the infimum in the definition of $C_{\mathbf{H}}'$ is achieved. To accomplish this a vector-valued form of the Clarkson inequalities is needed.

2.2.25 Lemma. Suppose $(\mathbf{M},\mathbf{F},\nu)$ is a measure space and H a real Hilbert space with norm $\|\ \|$. If f, g are H-valued functions on M with $\|f\|$, $\|g\|$, $\|f+g\|$, $\|f-g\|$, $\mathbf{F}$ measurable, then

$$\int \left\|\frac{f-g}{2}\right\|^p d\nu + \int \left\|\frac{f+g}{2}\right\|^p d\nu \le \frac{1}{2}\int \|f\|^p d\nu + \frac{1}{2}\int \|g\|^p d\nu \tag{2.2.26}$$

for $2 \le p < \infty$, and

$$\left(\int \left\|\frac{f-g}{2}\right\|^p d\nu\right)^{1/(p-1)} + \left(\int \left\|\frac{f+g}{2}\right\|^p d\nu\right)^{1/(p-1)} \tag{2.2.27}$$

$$\le \left(\frac{1}{2}\int \|f\|^p d\nu + \frac{1}{2}\int \|g\|^p d\nu\right)^{1/(p-1)}$$

for $1 < p \le 2$.

Proof. Given $u,v \in H$, let L be their span. Since the scalar field is $\mathbf{R}$, there is a linear map $\phi : L \to \mathbf{C}$ such that $\|x\| = |\phi(x)|$ for $x \in L$. Using 15.4 and 15.7 in [HS] it follows that

$$\left|\frac{\phi(u)-\phi(v)}{2}\right|^p + \left|\frac{\phi(u)-\phi(v)}{2}\right|^p \le \frac{1}{2}|\phi(u)|^p + \frac{1}{2}|\phi(v)|^p$$

for $2 \le p < \infty$, and

$$\left|\frac{\phi(u)-\phi(v)}{2}\right|^{p/(p-1)} + \left|\frac{\phi(u)-\phi(v)}{2}\right|^{p/(p-1)}$$

$$\le \left(\frac{1}{2}|\phi(u)|^p + \frac{1}{2}|\phi(v)|^p\right)^{1/(p-1)}$$

for $1 < p \le 2$.

Using linearity of ϕ and $\|x\| = |\phi(x)|$ and letting $u = f$, $v = g$ leads to

$$\left\|\frac{f+g}{2}\right\|^p + \left\|\frac{f-g}{2}\right\|^p \le \frac{1}{2}\|f\|^p + \frac{1}{2}\|g\|^p$$

for $2 \le p < \infty$, and

$$\left\|\frac{f+g}{2}\right\|^{p/(p-1)} + \left\|\frac{f-g}{2}\right\|^{p/(p-1)} < \left(\frac{1}{2}\|f\|^p + \frac{1}{2}\|g\|^p\right)^{1/(p-1)}$$

for $1 < p < 2$.

(2.2.26) follows by integration of the $2 < p < \infty$ inequality. Using Minkowski's inequality for powers between zero and one, 12.9 in [HS], it follows that

$$\left(\int \left\|\frac{f+g}{2}\right\|^p d\nu\right)^{1/(p-1)} + \left(\int \left\|\frac{f-g}{2}\right\|^p d\nu\right)^{1/(p-1)}$$

$$< \left(\int \left(\left\|\frac{f+g}{2}\right\|^{p/(p-1)} + \left\|\frac{f-g}{2}\right\|^{p/(p-1)}\right)^{p-1} d\nu\right)^{1/(p-1)}$$

$$\left(\int \left(\frac{1}{2}\|f\|^p + \frac{1}{2}\|g\|^p\right) d\nu\right)^{1/p}$$

from above, and so (2.2.27) holds. ∎

<u>2.2.28</u> <u>Proposition</u>. If $H = C_0^\infty(\Omega)$, $E \subseteq \Omega$, and the inequality

$$\int |\phi|^p d\omega < c \int |\nabla\phi|^p d\nu \qquad (2.2.29)$$

holds for all $\phi \in C_0^\infty(\Omega)$, then

$$C_H(E) = \inf\{\int |\nabla u|^p d\nu : u \in W_0^{1,p}(\Omega)\ , \qquad (2.2.30)$$

$$u > 1 \text{ on } E \text{ quasi-everywhere,}$$

$$\text{and } u \text{ is quasicontinuous}\}\ ,$$

and

$$C_H(\{|u| > \lambda\}) < \frac{2}{\lambda} \int |\nabla u|^p d\nu \qquad (2.2.31)$$

for all quasicontinuous $u \in W_0^{1,p}(\Omega)$.

(2.2.32) If $u_n \in W_0^{1,p}(\Omega)$ is quasicontinuous for $n = 1,2,\ldots$ and $\{u_n\}$ is Cauchy in $W_0^{1,p}(\Omega)$, then there exists $u \in W_0^{1,p}(\Omega)$, u quasicontinuous and a subsequence $\{n_i\}$ such that $u_{n_i} \to u$ in $W_0^{1,p}(\Omega)$ and uniformly off open sets of arbitrarily small capacity, and so pointwise quasi-everywhere as well.

(2.2.33) If $C_H(E) < \infty$, there exists $u \in W_0^{1,p}(\Omega)$ such that u is quasicontinuous, $0 < u < 1$, $u = 1$ everywhere on E, and $\int |\nabla u|^p d\nu = C_H(E)$.

(2.2.34) If $(u_1\ , \nabla u_1)$ is an extremal in the sense that $u_1 \in W_0^{1,p}(\Omega)$, u is quasicontinuous, $u_1 > 1$ quasi-everywhere on

E, and $\int |\nabla u_1|^p d\nu = C_H(E)$, then $(u_1, \nabla u_1) = (u, \nabla u)$ in $W_0^{1,p}(\Omega)$, where $(u, \nabla u)$ is as in 2.2.33. Also, $u = u_1$ quasi-everywhere.

(2.2.35) If Ω' is open and bounded, $\bar{\Omega}' \subseteq \Omega$, $E \subset \Omega'$, and $C_H(E) = 0$ for $H = C_0^\infty(\Omega)$, then $C_{H'}(E) = 0$ for $H' = C_0^\infty(\Omega')$. In consequence, if either u is bounded or $\nu \leq c\omega$ and if $u \in W_{loc}^{1,\infty}(\Omega)$, u locally quasicontinuous and $u|_{\Omega'} \in W_0^{1,p}(\Omega')$, then $u|_{\Omega'}$ is quasicontinuous with respect to Ω'.

Remarks. If $(u, \nabla u)$ is an extremal as in 2.2.34, then it will be called a capacitary extremal of E. It is clear that it is essentially unique.

The assumption of the Poincaré inequality 2.2.29 may be avoided by the use of a definition of $W_0^{1,p}(\Omega)$ which does not require $u \in L^p(\omega, \Omega)$ for $u \in W_0^{1,p}(\Omega)$.

Proof of Proposition 2.2.28. As will be shown in Proposition 2.2.41, the inequality (2.2.29) implies that if $H = C_0^\infty(\Omega)$, then $\omega(E) \leq c^p C_H(E)$ for all Borel sets E. Thus any set of C_H capacity zero is automatically of ω measure zero. Therefore if $u \in W_0^{1,p}(\Omega)$, then redefining it on a set of C_H capacity zero will not alter the $L^p(\omega, \Omega)$ equivalence class in which it lies and so it is unchanged as an element of $W_0^{1,p}(\Omega)$. Also, if it is quasicontinuous initially, then from the definition of quasicontinuity it will remain so. This property will be used periodically throughout the rest of the section.

Given $u \in W_0^{1,p}(\Omega)$, u quasicontinuous and $u \geq 1$ quasi-everywhere on E, pick $\phi_n \in C_0^\infty(\Omega)$ such that $\phi_n \to u$ in $W_0^{1,p}(\Omega)$ and uniformly pointwise off sets of arbitrarily small capacity. Let

$$E_{N,\delta} = \{x \in \Omega : \phi_n(x) \geq 1 - \delta \text{ for all } n \geq N\} ,$$

so

$$C_\lambda(E_{N,\delta}) \leq \frac{1}{(1-\delta)^p} \int |\nabla\phi|^p d\nu \leq \frac{1}{(1-\delta)^p} \left(\int |\nabla u|^p d\nu + \varepsilon_N\right) ,$$

where $\varepsilon_N \to 0$ as $N \to \infty$. Since the ϕ_n converge uniformly to u off sets of arbitrarily small capacity, there exist $F_N \subset \Omega$ such that $C_H(F_N) \leq \varepsilon_N'$, $\varepsilon_N' \to 0$ as $N \to \infty$, and $\{u \geq 1\} \subseteq E_{N,\delta} \cup F_N$. As a result,

$$C_H(E) \leq C_H(\{u \geq 1\}) \leq C_H(E_{N,\delta}) + C_H(F_N)$$

$$\leq \frac{1}{(1-\delta)^p} \left(\int |\nabla u|^p d\nu + \varepsilon_N\right) + \varepsilon_N' .$$

Let $N \to \infty$ and then $\delta \to 0$ to get $C_H(E) \le \int |\nabla u|^p d\nu$. Given $\varepsilon > 0$, pick O open, $E \subseteq O$ such that $C_H(O) \le C_H(E) + \varepsilon$. Choose K_n compact, $K_n \uparrow O$ such that $C(K_n) \uparrow C(O)$ and $\phi_n \in C_0^\infty(\Omega)$ such that $\phi_n \ge 1$ on K_n and $\int |\nabla \phi_n|^p d\nu \le C_H(K_n) + 2^{-n}$.

(2.2.36) By Lemma 2.2.25, using $H = \mathbb{R}^d$, it follows that

$$\int \left|\frac{\nabla\phi_n - \nabla\phi_m}{2}\right|^p d\nu + \int \left|\frac{\nabla\phi_n + \nabla\phi_m}{2}\right|^p d\nu$$

$$\le \frac{1}{2}\int |\nabla\phi_n|^p d\nu + \frac{1}{2}\int |\nabla\phi_m|^p d\nu \quad \text{for } p \ge 2 ,$$

and

$$\left(\int \left|\frac{\nabla\phi_n - \nabla\phi_m}{2}\right|^p d\nu\right)^{1/(p-1)} + \left(\int \left|\frac{\nabla\phi_n + \nabla\phi_m}{2}\right|^p d\nu\right)^{1/(p-1)}$$

$$\le \left(\frac{1}{2}\int |\nabla\phi_n|^p d\nu + \frac{1}{2}\int |\nabla\phi_m|^p d\nu\right)^{1/(p-1)} \quad \text{for } 1 < p \le 2 .$$

$\frac{\phi_n + \phi_m}{2} \ge 1$ on $K_{n\wedge m}$, where $n\wedge m = \min\{n,m\}$, so $C_H(K_{n\wedge m}) \le \int \left|\frac{\nabla\phi_n + \nabla\phi_m}{2}\right|^p d\nu$. Using this inequality on the above and then taking the $\limsup_{n,m\to\infty}$, recalling that $C(K_n) \to C(O)$, it follows that

$$\limsup_{n,m\to\infty} \int \left|\frac{\nabla\phi_n - \nabla\phi_m}{2}\right|^p d\nu + C_H(O) \le C_H(O)$$

for $p \ge 2$. A similar inequality holds for $1 < p \le 2$ so the $\nabla\phi_n$ are Cauchy in $\prod_{K=1}^{d} L^p(\nu,\Omega)$. The inequality (2.2.29) now implies that the ϕ_n are Cauchy in $L^p(\omega,\Omega)$ and so the ϕ_n are Cauchy in $W_0^{1,p}(\Omega)$. The ϕ_n converge to some $u \in W_0^{1,p}(\Omega)$ which can, by Proposition 2.2.21, be chosen such that a subsequence ϕ_{n_i} converges quasi-everywhere to u, and so $u \ge 1$ quasi-everywhere on O and u is quasicontinuous.

In addition,

$$\int |\nabla u|^p d\nu = \lim_{n\to\infty} \int |\nabla\phi_n|^p d\nu \le \lim_{n\to\infty} (C_H(K_n) + 2^{-n})$$

$$= C_H(O) \le C_H(E) + \varepsilon$$

and so (2.2.30) is established. ∎

Proof of (2.2.31). Using (2.2.30) and arguing as in Proposition 2.2.18, it follows that $C_H(\{|u| > \lambda\}) \leq \frac{2}{\lambda^p} \int |\nabla u|^p d\nu$ for all quasicontinuous $u \in W_0^{1,p}(\Omega)$. ∎

Proof of (2.2.32). Proceeding as in the first part of Proposition 2.2.19 implies that a subsequence of the u_n converges uniformly off sets of arbitrarily small capacity. Since the capacity of a set E can be approximated arbitrarily closely by capacities of open sets containing E, it follows that the exceptional sets above may be taken to be open. Let u be the $W_0^{1,p}(\Omega)$ limit of the u_n. Redefine it on a set of ω measure zero as in (2.2.22) so that $u_n \to u$ quasi-everywhere. Arguing as in the last part of (2.2.23) with ϕ_n a sequence in $C_0^\infty(\Omega)$ such that $\phi_n \to u$ in $W_0^{1,p}(\Omega)$, it follows that u is quasicontinuous. ∎

Proof of (2.2.33). Given $E \subseteq \Omega$, it can be seen from (2.2.30) that there exist quasicontinuous $u_n \in W_0^{1,p}(\Omega)$ such that $u_n \geq 1$ quasi-everywhere on E and

$$C_H(E) \leq \int |\nabla u_n|^p d\nu \leq C_H(E) + \varepsilon_n, \quad \varepsilon_n \to 0 \quad \text{as} \quad n \to \infty .$$

Using Clarkson's inequalities as before, it follows that the u_n are Cauchy and therefore by (2.2.32) there exists $u \in W_0^{1,p}(\Omega)$ quasi-continuous and a subsequence $\{n_i\}$ such that $u_{n_i} \to u$ in $W_0^{1,p}(\Omega)$ and pointwise quasi-everywhere, so $\int |\nabla u|^p d\nu = C_H(E)$ and $u \geq 1$ quasi-everywhere on E. Use 2.2.5 with $h_{0,1}$ to show that $(h_{0,1}(u), \chi_{\{0<u<1\}}\nabla u) \in W_0^{1,p}(\Omega)$. It is also clear from the proof of 2.2.5 that $h_{0,1}(u)$ is quasicontinuous. Also

$$\int |\chi_{\{0<u<1\}}\nabla u|^p d\nu \leq \int |\nabla u|^p d\nu \leq C_H(E) ,$$

but $C_H(E) \leq \int |\chi_{\{0<u<1\}}\nabla u|^p d\nu$ from (2.2.30), so equality holds and $h_{0,1}(u)$ is the required extremal after redefinition on a set of capacity zero. ∎

Proof of (2.2.34). Clarkson's inequalities imply that $\int |\nabla u_1 - \nabla u|^p d\nu = 0$ and the inequality $\int \phi^p d\omega \leq c \int |\nabla \phi|^p d\omega$, which holds for u_1, u by taking limits, then implies that $\int |u_1 - u|^p d\omega = 0$ and so $(u_1, \nabla u_1) = (u, \nabla u)$ in $W_0^{1,p}(\Omega)$. If $\phi_n \in C_0^\infty(\Omega)$ and $\phi_n \to (u, \nabla u)$ in $W_0^{1,p}(\Omega)$, then $\phi_n \to (u_1, \nabla u_1)$ in $W_0^{1,p}(\Omega)$, in which case there is a subsequence of the ϕ_n which converges to both u and u_1 quasi-everywhere so that $u = u_1$ quasi-everywhere. ∎

Proof of (2.2.35). If $K \subseteq \Omega'$ is compact and $C_{\mathbf{H}}(K) = 0$, then $\exists\ \phi_n \in C_0^\infty(\Omega)$ such that $\phi_n \geq 1$ on K and $\int |\nabla\phi_n|^p d\nu \to 0$. If

$$g(x) = \begin{cases} 1 & x \geq \frac{1}{2}\,, \\ 2x & -\frac{1}{2} \leq x \leq \frac{1}{2}\,, \\ -1 & x \leq -\frac{1}{2}\,, \end{cases}$$

$\sigma \in C_0^\infty(\mathbf{R})$, support $\sigma \subseteq [-\frac{1}{2}, \frac{1}{2}]$, $\sigma \geq 0$, $\int \sigma = 1$, and $\sigma(x) = \sigma(-x)$, then let $f = \sigma * g$ so that $|f| \leq 1$, $f(x) = 1$ if $x \geq 1$, $f(0) = 0$, and $|f'| \leq 2$. It now follows that $f(\phi_n) \in C_0^\infty(\Omega)$, $f(\phi_n) = 1$ on K and $\int |\nabla f(\phi_n)|^p d\nu \leq 2 \int |\nabla\phi|^p d\nu \to 0$. An application of Clarkson's inequalities implies that $f(\phi_n) \to 0$ in $W^{1,p}(\Omega)$ as in 2.2.36, and so $f(\phi_n) \to 0$, ω almost everywhere. Choose $\eta \in C_0^\infty(\Omega')$, $0 \leq \eta \leq 1$ and $\eta = 1$ on K so that $\eta f(\phi_n) = 1$ on K and

$$\int |\nabla(\eta f(\phi_n))|^p d\nu \leq \int |\nabla\eta|^p |f(\phi_n)|^p d\nu + 2 \int |\nabla\phi|^p d\nu \to 0$$

since $|f(\phi_n)| \leq 1$, $f(\phi_n) \to 0$, ω almost everywhere, and ν is absolutely continuous to ω. It is now clear that $C_{\mathbf{H}'}(K) = 0$. If $E \subseteq \Omega'$ and $C_{\mathbf{H}}(E) = 0$, then $C_{\mathbf{H}'}(K) = C_{\mathbf{H}}(K) = 0$ for all $K \subseteq E$, K compact but then the capacitability of $C_{\mathbf{H}'}$ as in Proposition 2.2.15 implies that $C_{\mathbf{H}'}(E) = 0$.

Suppose $u \in W^{1,p}_{loc}(\Omega)$, u locally quasicontinuous and $u|_{\Omega'} \in W_0^{1,p}(\Omega')$. Choose $\{\phi_n\}$, $\phi_n \in C^\infty(\Omega)$ such that $\phi_n \to u$ in $W^{1,p}_{loc}(\Omega)$ and pointwise quasi-everywhere and $\{\psi_n\}$, $\psi_n \in C_0^\infty(\Omega')$ such that $\psi_n \to u|_{\Omega'}$ in $W_0^{1,p}(\Omega')$ and pointwise quasi-everywhere. $\bar{\Omega}'$ is compact since Ω' is bounded so $\phi_n|_{\Omega'} \to u|_{\Omega'}$ in $W^{1,p}(\Omega')$ and from the above $\phi_n|_{\Omega'} \to u|_{\Omega'}$, Ω'-quasi-everywhere, but then $u|_{\Omega'}$ is Ω'-locally quasicontinuous. By (2.2.24) it now follows that $\psi_n \to u|_{\Omega'}$, Ω'-quasi-everywhere, and so $u|_{\Omega'}$ is Ω'-quasicontinuous. ∎

The set function $\bar{C}_{H,P}$ defined in Section 2.1.0 is unfortunately not subadditive, even in the case of Lebesgue measure, though it can be redefined, as $C_{H,P}$ has been, to give a more natural measure of non-compact sets. If Ω' is open, $\bar{\Omega}' \subseteq \Omega$, $\omega(\Omega) < \infty$, λ is a positive Borel measure with $\lambda(\Omega') = 1$ and $\bar{H}$ is a subset of $C_0^\infty(\Omega)|_{\Omega'}$ closed under addition, and under composition with $C^\infty(\mathbf{R})$ functions having bounded derivative, then let

$$\bar{C}'_{\bar{H}}(K) = \inf\{\int_{\Omega'} |\nabla\phi|^p d\nu : \phi \in \bar{H},\quad \phi \geq 1 \text{ on } K, \text{ and } \int_{\Omega'} \phi d\lambda = 0\}$$

for all sets K, $K = K' \cap \Omega'$ for some compact $K' \subseteq \Omega$,

$$\bar{C}'_{\bar{H}}(O) = \sup\{\bar{C}'_{\bar{H}}(K) : K = K' \cap \Omega' \text{ for some compact } K' \subseteq \Omega \text{ and } K \subseteq O\}$$

and

$$\bar{C}_{\bar{H}}(E) = \inf\{\bar{C}'_{\bar{H}}(O) : O \text{ is open, } E \subseteq O \subseteq \Omega'\}$$

for all $E \subseteq \Omega'$.

2.2.37 Proposition. $\bar{C}'_{\bar{H}}(E) = \bar{C}'_{\bar{H}}(E)$ if E is open or $E = K \cap \Omega'$ for some compact $K \subseteq \Omega$.

Proof. The proof is virtually identical to that of Proposition 2.2.14 but with compact sets replaced by the intersections with Ω' of compact subsets of Ω. ■

2.2.38 Proposition. Assume $\lambda \leq c\omega$, $\nu \leq c\omega$, and

$$\int |\phi - \int \phi d\lambda| \omega \leq c \int |\nabla\phi|^p d\nu \tag{2.2.39}$$

for all $\phi \in \bar{H}$.

If $K = K' \cap \Omega'$ with $K' \subseteq \Omega$ compact and $\bar{C}_{\bar{H}}(K) < \infty$, then $\exists\, u \in W^{1,p}(\Omega)$ such that u is locally quasicontinuous, $\int u d\lambda = 0$, $u \geq 1$ quasi-everywhere on K, and $\int |\nabla u|^p d\nu = \bar{C}_{\bar{H}}(K)$.

Proof. Choose $\phi_n \in \bar{H}$ such that $\phi_n \geq 1$ on K, $\int \phi_n d\lambda = 0$, and $\int |\nabla\phi_n|^p d\nu \to \bar{C}_{\bar{H}}(K)$. Use Clarkson's inequalities as in 2.2.36 to show that the $\nabla\phi_n$ are Cauchy in $\prod_{K=1}^{d} L^p(\nu,\Omega)$. Inequality (2.2.39) then implies that the ϕ_n are Cauchy in $L^p(\omega,\Omega)$ so the ϕ_n converge in $W^{1,p}(\Omega)$ to some u which can be chosen locally quasicontinuous by (2.2.24), in which case for some subsequence $\{n_i\}$ the $\phi_{n_i} \to u$ pointwise quasi-everywhere and $u \geq 1$ quasi-everywhere on K. Since $\lambda \leq c\omega$, it follows that the ϕ_n converge to u in $L^1(\lambda,\Omega)$ so that $\int u d\lambda = 0$. Finally, $\int |\nabla u|^p d\nu = \bar{C}_{\bar{H}}(K)$ since $\int |\nabla\phi_n|^p d\nu \to \bar{C}_{\bar{H}}(K)$. ■

Sobolev Inequalities. The characterizations of the weights for the two inequalities (2.1.4) and (2.1.15) will be translated into the present setting, (2.2.42) and (2.2.46), and weights of the form $\mathrm{dist}^\sigma(x,K)$ will be shown to be admissible.

A number of sufficient conditions for special cases of (2.2.42) and (2.2.46) appear in the literature. The condition on ω, ν

assumed in [K], [MS], and [T1] is fairly strict. A result of Muckenhoupt and Wheeden [MW] introduces a less strict condition but the resultant inequalities are not useful here, since they assume that the density of ν is a fixed power of the density of ω. Welland [W] has given a simplified proof of this result, the methods of which (along with the Besicovitch covering lemma [G]) can easily be adapted to yield suitable inequalities if ω, ν satisfy a condition of the form

$$\left(\int_B \omega(x)dx\right)^{1/q^*}\left(\int_B \nu(x)^{-1/(p^*-1)}dx\right)^{(p^*-1)/p^*} \leq c|B|^{(1-1/n)+\varepsilon}$$

for all balls $B \subseteq \Omega$, Ω a bounded open set and for some $\varepsilon > 0$, $p^* < p$, and $q^* = p^* \frac{q}{p}$. If $\omega = \nu$, then this includes the A_p weights [M1]. The assertion above will not be put into rigorous form and proven since it departs from the general direction of these notes. The fact that A_p weights are admissible for the appropriate Sobolev inequalities was also recognized by E. B. Fabes, C. E. Kenig and R. P. Serapioni [FKS] independently and at the same time as by the present author. The use of fractional integrals in this approach turns out to be too crude to allow a characterization of the weights needed for the Sobolev inequality (2.2.42) and (2.2.46), since it annihilates important geometric properties of certain classes of weights. A simple example is provided by the weights $\omega(x) = \nu(x) = |x|^\alpha$ in $\Omega = B(0,1)$. The condition assumed in [K] forces $-\min\{p,n\} < \alpha < \min\{p,n(p-1)\}$, while altering the methods of [MW] and [W] allows $-n < \alpha < (p-1)n$ and the present methods allow $-n < \alpha < \infty$.

It will be assumed in Lemma 2.2.40 and Theorem 2.2.41 that $H = C_0^\infty(\Omega)$ and $\bar{H} = C_0^\infty(\Omega)|_{\Omega'}$, where Ω' is open $\bar{\Omega}' \subseteq \Omega$.

<u>2.2.40 Lemma</u>. If $K \subseteq \Omega$ is compact, then $\exists\, K_n \subseteq \Omega$ such that $K_n = \{\phi_n \geq 1\}$ for some $\phi_n \in C_0^\infty(\Omega)$, $K \subseteq K_n$, ∂K_n is C^∞ and $C_H(K_n) \to C_H(K)$.

If $K \subseteq \Omega'$, $K = K' \cap \Omega$ for some compact set $K' \subseteq \Omega$, then $\exists\, K_n' \subseteq \Omega$ such that $K_n' = \{\phi_n' \geq 1\}$ for some $\phi_n' \in C_0^\infty(\Omega)$, $\partial K_n'$ is C^∞, $K \subseteq K_n'$, and $\bar{C}_{\bar{H}}(K_n' \cap \Omega') \to \bar{C}_{\bar{H}}(K)$.

<u>Proof</u>. Let $K \subseteq \Omega$ be compact. Given ε, $0 < \varepsilon < 1$, choose $\phi \in C_0^\infty(\Omega)$ such that $\phi \geq 1$ on K and $\int |\nabla\phi|^p d\nu < C_H(K) + \varepsilon$. By the Morse-Sard theorem $\{|\nabla\phi| = 0\} \cap \{\phi = t\} = \emptyset$ for almost all $t \in \mathbf{R}$. Choose one such t_0 with $1 - \varepsilon < t_0 < 1$ so that $\{\phi \geq t_0\}$ is a compact set with smooth boundary, $K \subseteq \{\phi \geq t_0\}$ and

$$C_H(\{\phi \geq t_0\}) \leq \frac{1}{1-\varepsilon} \int |\nabla\phi|^p d\nu \leq \frac{1}{1-\varepsilon} (C_H(K) + \varepsilon) .$$

Sequences K_n, ϕ_n can now be easily chosen.

If $K \subseteq \Omega'$, $K = K' \cap \Omega'$ for some compact set $K' \subseteq \Omega$, then argue as above with ϕ chosen so that $\phi \geq 1$ on K, $\int_{\Omega'} \phi d\lambda = 0$ and $\int_{\Omega'} |\nabla\phi|^p d\nu \leq \bar{C}_{\bar{H}}(K) + \varepsilon$, and with $C_H(\{\phi \geq t_0\})$ replaced by $\bar{C}_{\bar{H}}(\{\phi \geq t_0\} \cap \Omega')$. ∎

2.2.41 Theorem. If $1 \leq p \leq q < \infty$, then

$$(\int |u|^q d\omega)^{1/q} \leq c_1(\int |\nabla u|^p d\nu)^{1/p} \tag{2.2.42}$$

for some $c_1 > 0$ and all $u \in C_0^\infty(\Omega)$ iff

$$\omega^{1/q}(K) \leq b_1 C_H^{1/p}(K) \tag{2.2.43}$$

for some $b_1 > 0$ and all compact sets $K \subseteq \Omega$ with C^∞ boundary iff

$$\omega^{1/q}(E) \leq b_1 C_H^{1/p}(E) \tag{2.2.44}$$

for some $b_1 > 0$ and all Borel measurable sets $E \subseteq \Omega$.

If $p = 1$ and ν is absolutely continuous with respect to Lebesgue measure on Ω with density $\bar{\nu} \in L^1(\Omega)$, then (2.2.42) holds iff

$$\omega^{1/q}(K) \leq b_1 \liminf_{\delta\to 0} \frac{1}{\delta} \int_{C_\delta} \bar{\nu} dx \tag{2.2.45}$$

for some $b_1 > 0$ and all compact sets K with C^∞ boundary, where $C_\delta = \{x \in \Omega - K : \operatorname{dist}(x,K) < \delta\}$.

If $\bar{\nu}$ is continuous, then this reduces to $\omega^{1/q}(K) \leq b_1 \int_{\partial K} \bar{\nu} dH^{n-1}$.

If $\omega(\Omega') < \infty$, $\lambda(\Omega') = 1$ and $1 \leq p \leq q < \infty$, then

$$(\int_{\Omega'} |u - \int_{\Omega'} u d\lambda|^q d\omega)^{1/q} \leq c_2(\int_{\Omega'} |\nabla u|^p d\nu)^{1/p} \tag{2.2.46}$$

for some $c_2 > 0$ and for all $u \in C_0^\infty(\Omega)|_{\Omega'}$, iff

$$\omega^{1/q}(K) \leq b_2 \bar{C}_{\bar{H}}^{1/p}(K) \tag{2.2.47}$$

for some $b_2 > 0$ and all $K \subseteq \Omega$ such that $K = K' \cap \Omega$ for some compact set $K' \subseteq \Omega$, iff

$$\omega^{1/q}(E) \leq b_2 \bar{c}_{\bar{H}}^{1/p}(E) \tag{2.2.48}$$

for some $b_2 > 0$ and all Borel measurable sets $E \subseteq \Omega'$.

If $p = 1$ and ν is absolutely continuous with density $\bar{\nu} \in L^1(\Omega')$, then (2.2.46) holds iff

$$\omega^{1/q}(K)\lambda(\Omega' - K) \leq b_2 \limsup_{\delta \to 0} \frac{1}{\delta} \int_{C'_\delta} \bar{\nu}dx \tag{2.2.49}$$

for some $b_2 > 0$ and all $K \subseteq \Omega'$ such that $K = K' \cap \Omega'$ for some compact set $K' \subseteq \Omega$, and where $C'_\delta = \{x \in \Omega' - K : \text{dist}(x,K) < \delta\}$.

If c_i, b_i, $i = 1,2$, are chosen as small as possible, then $b_1 \leq c_1 \leq p^{1/q} p'^{1/p'} b_1$ and $b_2 \leq c_2 \leq 2p^{1/q} p'^{1/p'} b_2$.

Proof. Consider Theorem 2.1.7. It is claimed that (2.1.9), (2.2.43), and (2.2.44) are all equivalent. It is clear that (2.2.44) implies both (2.1.9) and (2.2.43). (2.2.44) follows from each of these in a similar manner so only one implication will be done explicitly. Assume (2.1.9). Given a compact set $K \subseteq \Omega$, Lemma 2.2.40 supplies a sequence $\{K_n\}$ of compact sets of the type considered in 2.2.40 such that $\omega^{1/q}(K) \leq \omega^{1/q}(K_n) \leq b_1 c_H^{1/p}(K_n) \to b_1 c_{\mathbf{H}}^{1/p}(K)$. Given a Borel set $E \subseteq \Omega$, use the regularity of ω to choose a sequence of compact sets K'_n such that $K'_n \subseteq E$ and $\omega(K'_n) \to \omega(E)$, so

$$\omega^{1/q}(E) = \lim_{n\to\infty} \omega^{1/q}(K_n) \leq b_1 \limsup_{n\to\infty} c_H(K_n) \leq b_1 c_{\mathbf{H}}(E) ,$$

and (2.2.44) is verified.

The equivalence of (2.2.42) and (2.2.45) now follows directly from Theorem 2.1.22.

The second half of the theorem follows in a similar manner to the first using Theorem 2.1.17 instead of 2.1.7. ∎

It will be shown in Theorem 2.2.56 that weights of the form $\text{dist}^\alpha(x,K)$ admit Sobolev inequalities of the type (2.2.42) and (2.2.46). These will be used in Chapter 3 to demonstrate the Hölder continuity of solutions of certain differential equations which have these weights as degeneracies.

It will first be shown that two weighted isoperimetric inequalities hold under the conditions (2.2.51), (2.2.52), and (2.2.53). The rather technical verification of these conditions for specific geometries is left to the proof of Theorem 2.2.56.

Let ω, ν be nonnegative Borel functions defined everywhere on $\bar{B}(x_0,2R_0) \subseteq \mathbf{R}^d$, $d \geq 2$. For each r, $0 < r < 2R_0$, let C_r and

D_r be Borel measurable subsets of $\bar{B}(x_0,2R_0)$, C_r will correspond to sets where ν is "small" and D_r to sets where ω is "large". Finally, let P_z be the projection of $\mathbf{R}^d$ onto the hyperplane $\{x \in \mathbf{R}^d : x\cdot z = 0\}$, $z \in \mathbf{R}^d$, $z \neq 0$, $P_z(\mathbf{R}^d)$ will sometimes be casually identified with $\mathbf{R}^{d-1}$. $a(d)$ will be the d-dimensional measure of the unit ball in $\mathbf{R}^d$.

2.2.50 Proposition. Assume that the following conditions hold for all $B(x,r) \subseteq B(x_0,2R_0)$ and $z \in \mathbf{R}^d$, $z \neq 0$.

$$H^{d-1}(P_z(C_r \cap B(x,r))) < \frac{a(d)}{2^{5d+4}}\, r^{d-1}\ , \tag{2.2.51}$$

$$\omega(B(x,r)) < c_1\omega(B(x,r) - D_r) \tag{2.2.52}$$

$$\max_{B(x,r)-D_r} \omega < c_2|B(x,r)|^{((d-1)/d)-1/q} \min_{B(x,r)-C_r} \nu \tag{2.2.53}$$

for some $q > 1$ and c_1, c_2 independent of r, x, z, then there exists a constant $c(d)$ such that

(2.2.54) If X is open and $|B(x_0,R_0) - X| > \frac{1}{2}|B(x_0,R_0)|$, then

$$\omega^{1/q}(X \cap B(x_0,R_0)) < c(d)c_1^{1/q}c_2\nu_{d-1}(\partial X \cap B(x_0,R_0)).$$

(2.2.55) If X is open and $X \subseteq B(x_0,R_0)$, then

$$\omega^{1/q}(X) < c(d)c_1^{1/q}c_2\nu_{d-1}(\partial X)\ ,$$

where $\omega(E) = \int_E \omega$ and $\nu_{d-1}(E) = \int_E \nu dH^{d-1}$.

The abbreviations max, min have been used instead of sup and inf to emphasize that it is the true supremum or infimum which is indicated and not the essential supremum or infimum. The proof of Proposition 2.2.50 will be deferred till later.

Suppose $K \subseteq \mathbf{R}^d$, $|\bar{K}| = 0$, $d > 2$, and $\alpha,\beta \in \mathbf{R}$. Then let $\omega(x) = \text{dist}^{\alpha}(x,K)$, $\nu(x) = \text{dist}^{\beta}(x,K)$, and $A(t) = \{x \in \mathbf{R}^d : \text{dist}(x,K) < t\}$.

2.2.56 Theorem. If $1 < q < \frac{d}{d-1}$, $\frac{\alpha + d}{q} > d + \beta - 1$, and the following conditions hold for all $B(x,r) \subseteq B(x_0,R_0)$ and $z \in \mathbf{R}^d$, $z \neq 0$:

(2.2.57) If $\beta > 0$, then $H^{d-1}(P_z(B(x,r) \cap A(\varepsilon_1 r))) < \frac{a(d)}{2^{5d+4}}\, r^{d-1}$ for some ε_1, $0 < \varepsilon_1 < \frac{1}{2}$.

(2.2.58) If $\alpha < 0$, then $\omega(B(x,r)) \leq c_1\omega(B(x,r) - A(\varepsilon_2 r))$ for some $c_1 > 0$ and some ε_2, $0 < \varepsilon_2 < \frac{1}{2}$,

with ε_1, ε_2, and c_1 independent of r, R_0, x, z, x_0, then there exists a constant c_2 independent of x_0, R_0 such that

(2.2.59) If $X \subseteq \mathbf{R}^d$, then

$$\omega^{1/q}(X \cap B(x_0,R_0))\omega(B(x_0,R_0) - X)$$

$$\leq c_2 R_K^{(\alpha/q)-\beta} R_0^{(d/q)-d+1} \omega(B(x_0,R_0))\nu_{d-1}(\partial X \cap B(x_0,R_0)).$$

(2.2.60) If $X \subseteq B(x_0,R_0)$, then

$$\omega^{1/q}(X) \leq c_2 R^{(\alpha/q)-\beta} R_0^{(d/q)-d+1} \nu_{d-1}(\partial X) ,$$

where $R_K = \max\{R_0 , \mathrm{dist}(x_0,K)\}$.

In addition, if $1 > \frac{1}{p} > 1 - \frac{1}{q}$, $\frac{1}{t} = \frac{1}{p} + \frac{1}{q} - 1$ and $\sigma(x) = \mathrm{dist}^{p\beta-(p-1)\alpha}(x,K) = \nu(x)^p\omega^{-(p-1)}$, then there exists a constant c_3 independent of x_0, R_0 such that

$$(\int |\phi|^t\omega)^{1/t} \leq c_3 R_K^{(\alpha/q)-\beta} R_0^{(d/q)-d+1} (\int |\nabla\phi|^p\sigma)^{1/p} \tag{2.2.61}$$

for all $\phi \in C_0^\infty(B(x_0,R_0))$,

$$(\int_B |\phi - \frac{1}{\omega(B)} \int_B \phi\omega|^t\omega)^{1/t} \leq c_3 R_K^{(\alpha/q)-\beta} R_0^{(d/q)-d+1} (\int_B |\nabla\varepsilon|^p\sigma)^{1/p} \tag{2.2.62}$$

for all $\phi \in C^\infty(B(x_0,2R_0))|_{B(0,R_0)}$ and $B = B(x_0,R_0)$.

If $K = \cup_{i=1}^n M_i$, M_i a compact C^2 manifold of co-dimension $\gamma_i \geq 2$, or a point ($\gamma_i = d$), and $\alpha > -\gamma_i$, $i = 1,\ldots,n$, then (2.2.57) and (2.2.58) are satisfied for all $x \in \mathbf{R}^d$ and $r \in (0,\infty)$ so that (2.2.61) and (2.2.62) hold for all x_0, R_0. In addtion, if $\beta p - (p - 1)\alpha > -\gamma$, then

$$R_K^{(\alpha/q)-\beta} R_0^{(d/q)-d+1} \leq c_4 \omega^{1/t}(B(x_0,R_0))\sigma^{-1/p}(B(x_0,R_0))R_0 \tag{2.2.63}$$

for some c_4 independent of x_0, R_0.

<u>Remarks</u>. The conditions on ω, σ allow σ/ω to degenerate to zero on K and also q may be chosen arbitrarily close to 1 so as to allow consideration of arbitrarily large p.

The assumption that M_i is a compact manifold is not necessary but merely convenient. The conclusions of the theorem are true for much more general sets K.

The specific estimate for the coefficient $R_K^{(\alpha/q)-\beta} R_0^{(d/q)-d+1}$ given in (2.2.63) is important since its existence will lead to a proof that solutions of certain differential equations with degeneracies of the form $\text{dist}^\sigma(x,K)$ are Hölder-continuous.

It is first convenient to prove a lemma which is a generalization of a lemma of Federer [F1] which he used to provide a simple proof of an important result of Gustin [GU].

__2.2.64 Lemma.__ If A, B are compact sets, $A \cup B$ convex with diameter δ and E is a Borel set, then

$$\frac{|A|}{\delta^d} \frac{|B|}{\delta^d} \leq \frac{a(d)}{\delta^{d-1}} H^{d-1}((A \cap B) - E) + \frac{1}{\delta^{2d-1}} \int_{|z| < \delta} H^{d-1}(P_z(E)) dz \ .$$

__Proof of Lemma 2.2.64.__ If A, B are compact sets, $A \cup B$ convex with diameter δ and E is a Borel measurable set, then

$$|A| \ |B| = \iint \chi_A(x) \chi_B(y) dx \ dy$$

$$= \iint \chi_A(x) \chi_B(x + z) dx \ dy$$

$$= \int_{|z| < \delta} |\{x : x \in A, \ x + z \in B\}| dz$$

$$\leq \int_{|z| < \delta} \int_{P_z(A \cap B)} H^1(\{x : x \in A \text{ and } x = \xi + tz \text{ for some } t\}) d\xi \ dz$$

$$\leq \delta \int_{|z| < \delta} H^{d-1}(P_z(A \cap B)) dz$$

$$\leq \delta \int_{|z| < \delta} H^{d-1}(P_z(((A \cap B) - E) \cup E)) dz$$

$$= \delta \int_{|z| < \delta} H^{d-1}(P_z((A \cap B) - E) \cup P_z(E)) dz$$

$$\leq \delta \int_{|z| < \delta} [H^{d-1}(P_z((A \cap B) - E)) + H^{d-1}(P_z(E))] dz$$

$$\leq a(d) \delta^{d+1} H^{d-1}((A \cap B) - E) + \delta \int_{|z| < \delta} H^{d-1}(P_z(E)) dz$$

Dividing by δ^{2d} now gives the result. ∎

__Proof of Proposition 2.2.50.__ Assume that (2.2.51), (2.2.52), and (2.2.53) hold and that X is open and $|B(x_0,R_0) - X| > \frac{1}{2} |B(x_0,R_0)|$.

__2.2.65.__ Given $x \in X \cap B(x_0,R_0)$, $\exists \ r$, $0 < r \leq 2R_0$, such that $|B(x,r) \cap B(x_0,R_0) \cap X| = \frac{1}{2} |B(x,r) \cap B(x_0,R_0)|$ since

$|B(x,r) \cap B(x_0,R_0) \cap X|$ is continuous in r, $|B(x,r) \cap B(x_0,R_0) \cap X| = |B(x,r)| = |B(x,r) \cap B(x_0,R_0)|$ for small r and

$$
\begin{aligned}
|B(x,r) \cap B(x_0,R_0) \cap X| &= |B(x_0,R_0) \cap X| \\
&= |B(x_0,R_0)| - |B(x_0,R_0) - X| \\
&< \tfrac{1}{2}\,|B(x_0,R_0)| = \tfrac{1}{2}\,|B(x,r) \cap B(x_0,R_0)|
\end{aligned}
$$

for $r = 2R_0$.

Let $r' = r(1-\varepsilon)$ and $R_0' = R_0(1-\varepsilon)$ for small ε so that $|B(x,r') \cap B(x_0,R_0') \cap X| = \frac{1}{2} f(\varepsilon)|B(x,r') \cap B(x_0,R_0')|$, where $f(\varepsilon) \to 1$ as $\varepsilon \to 0$. Now apply (2.2.51) and Lemma 2.2.64 with $A = \overline{B(x,r')} \cap \overline{B(x_0,R_0')} \cap \bar{X}$, $B = \overline{(B(x,r')} \cap \overline{B(x_0,R_0'))} - X$ and $E = C_r \cap \overline{B(x,r')}$ to get that

$$
\frac{|B(x,r') \cap B(x_0,R_0') \cap X|}{\delta^d}\;\frac{|(B(x,r') \cap B(x_0,R_0')) - X|}{\delta^d}
$$

$$
< \frac{a(d)}{\delta^{d-1}}\, H^{d-1}((\partial X \cap \overline{B(x,r')} \cap \overline{B(x_0,R_0')}) - C_r) + \frac{a^2(d)}{2^{5d+4}}\,\frac{r^{d-1}}{\delta^{d-1}}
$$

for $\delta = \operatorname{diam} \overline{B(x,r')} \cap \overline{B(x_0,R_0')}$.

A simple calculation shows that $\delta < 2r$ and

$$
|B(x,r) \cap B(x_0,R_0)| > a(d)\left(\frac{\min(r,R_0)}{2}\right)^d > a(d)\left(\frac{r}{4}\right)^d ,
$$

so letting $\varepsilon \to 0$ it follows that

$$
\frac{|B(x,r) \cap B(x_0,R_0) \cap X|}{|B(x,r) \cap B(x_0,R_0)|}\;\frac{|(B(x,r) \cap B(x_0,R_0)) - X|}{|B(x,r) \cap B(x_0,R_0)|}
$$

$$
< \frac{2^{5d+1}}{a(d)r^{d-1}}\, H^{d-1}((\partial X \cap B(x,r) \cap B(x_0,R_0)) - C_r) + \frac{1}{2^3} ,
$$

and by 2.2.65, $\frac{1}{8} < \frac{2^{5d+1}}{a(d)r^{d-1}} H^{d-1}((\partial X \cap B(x,r) \cap B(x_0,R_0) - C_r)$, and $|B(x,r)|^{(d-1)/d} < 2^{5d+1}a(d)^{-1/d}H^{d-1}((\partial X \cap B(x,r) \cap B(x_0,R_0)) - C_r)$. Using (2.2.52) and (2.2.53), it follows that

$$
\omega^{1/q}(B(x,r)) < c_1^{1/q}\omega^{1/q}(B(x,r) - D_r)
$$

$$
< c_1^{1/q} \max_{B(x,r)-D_r} |B(x,r)|^{1/q} < c_1^{1/q} c_2 \min_{B(x,r)-C_r} \nu\, |B(x,r)|^{(d-1)/d}
$$

$$< 2^{5d+4}a(d)^{-1/d}c_1^{1/q}c_2 \min_{B(x,r)-C_r} \nu H^{d-1}((\partial X \cap B(x,r) \cap B(x_0,R_0)) - C_r)$$

$$< 2^{5d+1}a(d)^{-1/d}c_1^{1/q}c_2\nu_{d-1}(\partial X \cap B(x,r) \cap B(x_0,R_0)) .$$

Now apply the Besicovitch covering lemma [G] to find $\mathbf{F}_i$, $i = 1,\dots,m$, each $\mathbf{F}_i$ a collection of pairwise disjoint closed balls $\bar{B}$, $B = B(x,r)$ as above, such that $\bigcup_i \mathbf{F}_i$ is a cover of $X \cap B(x_0,R_0)$. Since m can be chosen to be dependent only on d, it follows that $\omega(X \cap B(x_0,R_0)) < m\omega(\bigcup_{\bar{B}\in\mathbf{F}_i} \bar{B})$ for some i.

Let $\mathbf{F}_i = \{B(x_j,r_j)\}_j$ so that

$$\omega^{1/q}(X \cap B(x_0,R_0)) < m^{1/q}(\sum_j \omega(B(x_j,r_j)))^{1/q}$$

$$< m^{1/q} \sum_j \omega^{1/q}(B(x_j,r_j))$$

$$< m^{1/q}2^{5d+4}a(d)^{-1/d}c_1^{1/q}c_2 \sum_j \nu_{d-1}(\partial X \cap B(x_j,r_j) \cap B(x_0,R_0))$$

$$< c(d)c_1^{1/q}c_2\nu_{d-1}(\partial X \cap B(x_0,R_0))$$

since the balls $B(x_j,r_j)$ are pairwise disjoint.

(2.2.55) follows almost identically since given $x \in X$, there exists an r, $0 < r < 2R_0$, such that $|B(x,r) \cap X| = \frac{1}{2}|B(x,r)|$. This is true since $B(x,r) \subseteq X$ for small r, and $X \subseteq B(x,r)$ and $|B(x,r) \cap X| = |X| < |B(x_0,R_0)| < \frac{1}{2}|B(x,r)|$ for $r = 2R_0$. Lemma 2.2.64 is applied with $A = \overline{B(x,r')} \cap \bar{X}$ and $B = \overline{B(x,r')} - X$. A short calculation simpler than the one above then leads to

$$|B(x,r)|^{(d-1)/d} < 2^{d+4}a(d)^{-1/d}H^{d-1}((\partial X \cap B(x,r)) - C_r) ,$$

and the proof is concluded as above with the exception that $B(x_0,R_0)$ does not appear. ∎

Proof of Theorem 2.2.56. Recall that $A(t)=\{x \in \mathbb{R}^d : \operatorname{dist}(x,K) < t\}$. Let $C_r = A(\varepsilon_1 r)$ if $\beta > 0$ and $C_r = \emptyset$ if $\beta < 0$, and let $D_r = A(\varepsilon_2 r)$ if $\alpha < 0$ and $D_r = \emptyset$ if $\alpha > 0$. Assumptions (2.2.51) and (2.2.52) now follow from (2.2.57) and (2.2.58). (2.2.53) is verified as follows.

There are a number of cases to consider, depending on the relative geometry of K and $B(x,r)$ and the sign of α and β. Assume $B(x,r) \subseteq B(x_0,R_0)$, and let $r_1 = \operatorname{dist}(x,K)$.

A. If $r < \frac{1}{2} r_1$, then $A(\varepsilon_i r) \cap B(x,r) = \emptyset$, $i = 1,2$, since $\varepsilon_i < \frac{1}{2}$, and so

$$\max_{B(x,r)-D_r} \omega < \begin{cases} (\frac{3}{2} r_1)^\alpha & \alpha > 0 \\ (\frac{r_1}{2})^\alpha & \alpha < 0 , \end{cases}$$

$$\min_{B(x,r)-C_r} \nu > \begin{cases} (\frac{3r_1}{2})^\beta & \beta < 0 \\ (\frac{r_1}{2})^\beta & \beta > 0 . \end{cases}$$

B. If $r > \frac{r_1}{2}$,

$$\max_{B(x,r)-D_r} \omega < \begin{cases} (3r)^\alpha & \alpha > 0 \\ (\varepsilon_2 r)^\alpha & \alpha < 0 , \end{cases}$$

$$\min_{B(x,r)-C_r} \nu > \begin{cases} (3r)^\beta & \beta < 0 \\ (\varepsilon_1 r)^\beta & \beta > 0 . \end{cases}$$

The proof of (2.2.53) is virtually identical in each of the cases so only one will be done explicitly. If $\alpha < 0$, $\beta > 0$, and $r < \frac{r_1}{2}$, then

$$\begin{aligned} \max_{B(x,r)-D_r}{}^{1/q} \omega &< (\tfrac{r_1}{2})^{\alpha/q} < (\tfrac{r_1}{2})^{\alpha/q} (\tfrac{r_1}{2})^{-\beta} \min_{B(x,r)-C_r} \nu \\ &< 2^{\beta-(\alpha/q)} a(d)^{(1/q)-((d-1)/d)} r_1^{(\alpha/q)-\beta} r^{(d/q)-d+1} \\ &\qquad |B(x,r)|^{((d-1)/d)-1/q} \min_{B(x,r)-C_r} \nu \\ &< cR_K^{(\alpha/q)-\beta} R_0^{(d/q)-d+1} |B(x,r)|^{((d-1)/d)-1/q} \min_{B(x,r)-C_r} \nu \end{aligned}$$

for $R_K = \max\{R_0, \operatorname{dist}(x,K)\}$.

Since (2.2.51), (2.2.52), and (2.2.53) are verified, it follows that (2.2.54) and (2.2.55) hold. It will now be shown that the assumption that X is open is superfluous. If $\partial X \cap B(x_0,R_0)$ has positive d-dimensional measure, then (2.2.59) holds since $|\bar{K}| = 0$. Otherwise assume $|\partial X \cap B(x_0,R_0)| = 0$. Let $X' = \text{interior } X$ in which case $\partial X' \subseteq \partial X$ and $|B(x_0,R_0) - X| = |B(x_0,R_0) - X'|$, so if $|B(x_0,R_0) - X| \geq \frac{1}{2}|B(x_0,R_0)|$, then

$$\omega^{1/q}(X \cap B(x_0,R_0)) = \omega^{1/q}(X' \cap B(x_0,R_0)) \tag{2.2.66}$$

$$\leq cR_K^{(\alpha/q)-\beta}R_0^{(d/q)-d+1}\nu_{d-1}(\partial X' \cap B(x_0,R_0))$$

$$\leq cR_K^{(\alpha/q)-\beta}R_0^{(d/q)-d+1}\nu_{d-1}(\partial X \cap B(x_0,R_0)) .$$

A similar argument works for (2.2.55), in which case (2.2.60) is proven.

(2.2.59) follows directly from (2.2.66) if $|B(x_0,R_0) - X| \geq \frac{1}{2}|B(x_0,R_0)|$ since $\omega(B(x_0,R_0) - X) \leq \omega(B(x_0,R_0))$. If $|B(x_0,R_0) - X| \leq \frac{1}{2}|B(x_0,R_0)|$, then let $X' = B(x_0,R_0) - X$ so $|B(x_0,R_0) - X'| \geq \frac{1}{2}|B(x_0,R_0)|$, and (2.2.66) implies that

$$\omega^{1/q}(X' \cap B(x_0,R_0)) \leq cR_K^{(\alpha/q)-\beta}R^{(d/q)-d+1}\nu_{d-1}(\partial X' \cap B(x_0,R_0)) .$$

$\partial X' \subseteq \partial X$ so

$$\omega^{1/q}(X \cap B(x_0,R_0))\omega(B(x_0,R_0) - X) \leq \omega(B(x_0,R_0))\omega^{1/q}(B(x_0,R_0) - X)$$

$$\leq cR_K^{(\alpha/q)-\beta}R_0^{(d/q)-d+1}\omega(B(X_0,R_0))\nu_{d-1}(\partial X \cap B(x_0,R_0)) .$$

$\omega(B(x_0,R_0)) > 0$ since $|\bar{K}| = 0$, so let $\lambda = \omega/\omega(B(x_0,R_0))$. If ν were integrable, then (2.2.61) and (2.2.62) could be proven by appealing to 2.2.41, but there are interesting cases when this is not the case. To handle these, Section 2.1.0 is used in conjunction with a direct proof of the capacitary conditions involved. It will first be proven that $\omega^{1/q}(A) \leq c_2R_K^{(\alpha/q)-\beta}R_0^{(d/q)-d+1}\bar{K}_{H,1}(A)$ for all level sets $A = \{\phi \leq 0\}$,

$$\phi \in H = C^\infty(B(x_0,2R_0))|_{B(x_0,R_0)} \cap \{\phi : |\nabla\phi| \in L^p(\nu)\} ,$$

in which case it will follow from Section 2.1.0 and 1.3.5 that (2.2.62) is true in the case $p = 1$ and $t = q$. It will then be shown that (2.2.62) holds in general.

Given $A = \{\phi \leq 0\}$ for some $\phi \in H$, pick $\psi \in H$ with $\psi \leq 0$ on A.

$$\omega^{1/q}(A)\lambda(\{\psi > t\}) \leq \omega^{1/q}(\{\psi \leq t\})\lambda(B(x_0,R_0) - \{\psi \leq t\})$$

$$\leq c_2R_K^{(\alpha/q)-\beta}R_0^{(d/q)-d+1}\int_{\partial\{\psi<t\}\cap B(x_0,R_0)}\nu dH^{d-1}$$

for all $t > 0$, recalling that the domain of ψ is $B(x_0,R_0)$. It is claimed that $\bar{\mu}^*_\phi(t) = \int_{\partial\{\psi<t\}\cap B(x_0,R_0)} \nu dH^{n-1}$ a.e., so using the infinite arithmetic conventions in 1.3.5 it is seen that

$$\omega^{1/q}(A) \leq c_2 R_K^{(\alpha/q)-\beta} R_0^{(d/q)-d+1} \inf_{t\in(0,\infty)} \frac{\bar{\mu}^*_\psi(t)}{\lambda(\{\psi > t\})} .$$

Taking the infimum over all such ψ gives

$$\omega^{1/q}(A) \leq c_2 R_K^{(\alpha/q)-\beta} R_0^{(d/q)-d+1} \bar{K}_{H,1}(A) .$$

Theorem 1.3.5 states that $\bar{K}_{H,1}(A) = \bar{C}_{H,1}(A)$, so (2.2.61) holds by Theorem 2.1.17.

To prove the claim use the co-area formula to get that

$$\mu^*_\psi(E) = \int \chi_{\psi^{-1}(E)\cap B(x_0,R_0)} |\nabla\psi| \nu$$

$$= \int_E \int_{\partial\{\psi<t\}\cap B(x_0,R_0)} \nu dH^{d-1} dt$$

for $E \subseteq \mathbf{R}'$. These integrals are defined and finite since $|\nabla\psi| \in L^p(\nu)$ and ν can be realized by a pointwise everywhere, monotone increasing limit of bounded functions. This means that $\int_{\partial\{\psi<t\}\cap B(x_0,R_0)} \nu dH^{d-1}$ is locally integrable so that $\bar{\mu}^*_\psi(t) = \int_{\partial\{\psi<t\}\cap B(x_0,R_0)} \nu dH^{d-1}$ a.e. as required.

A similar proof, using the capacity $K_{H,1}$, establishes (2.2.61) for the case $p = 1$ and $t = q$.

To prove (2.2.62) in general it is necessary to prove the $t = q$, $p = 1$ case for ϕ replaced by $(\phi^+)^\alpha = \chi_{\{\phi>0\}}\phi^\alpha$ for $\alpha > 1$. To do this choose a C^∞ smoothing $\{f_n\}$ of $f(x) = \chi_{\{x>0\}}x$ such that $f_n \to f$ uniformly and $f'_n \to \chi_{\{x>0\}}$ with $0 \leq f'_n \leq \chi_{\{x>0\}}$. It is clear that $f^\alpha_n(\phi) \to (\phi^+)^\alpha$ in $L^1(\omega,\Omega)$ and $L^q(\omega,\Omega)$ so that substituting $f^\alpha_n(\phi)$ in (2.2.62) with $t = q$, $p = 1$, and then taking limits, it follows that (2.2.62) holds with $(\phi^+)^\alpha$ instead of ϕ. The same is true for $(\phi^-)^\alpha = \chi_{\{\phi>0\}}\phi^\alpha$.

One additional calculation is necessary. If a and b are non-negative and $\alpha > 1$, then $|a - b|^\alpha \leq |a^\alpha - b^\alpha|$ since if a is the largest of the two, then $a = \theta b$ for some θ, $0 \leq \theta \leq 1$, and $|a - b|^\alpha = (1 - \theta)^\alpha a^\alpha \leq (1 - \theta^\alpha)a^\alpha = |a^\alpha - b^\alpha|$. It will also be used, for $B = B(x_0,R_0)$, that

$$\Big(\frac{1}{\omega(B)}\int_B |u(x) - \frac{1}{\omega(B)}\int_B u(y)\omega(y)dy|^S\omega(x)dx\Big)^{1/S}$$

$$< \Big(\frac{1}{\omega(B)}\int_B \Big(\frac{1}{\omega(B)}\int_B |u(x) - u(y)|\omega(y)dy\Big)^S\omega(x)dx\Big)^{1/S}$$

$$< \Big(\frac{1}{\omega(B)}\int_B \Big(\frac{1}{\omega(B)}\int_B |u(x) - \frac{1}{\omega(B)}\int_B u(z)\omega(z)dz|\omega(y)dy\Big)^S\omega(x)dx\Big)^{1/S}$$

$$+ \Big(\frac{1}{\omega(B)}\int_B \Big(\frac{1}{\omega(B)}\int_B |u(y) - \frac{1}{\omega(B)}\int_B u(z)\omega(z)dz|\omega(y)dy\Big)^S\omega(x)dx\Big)^{1/S}$$

$$= 2\Big(\frac{1}{\omega(B)}\int_B |u(x) - \frac{1}{\omega(B)}\int_B u(z)\omega(z)dz|^S\omega(x)dx\Big)^{1/S}$$

so that the first two integrals are comparable. Now let $\alpha = \frac{t}{q} > 1$ and let $u(x) = \phi(x) - \frac{1}{\omega(B)}\int_B \phi(y)\omega(y)dy$ so that

$$\Big(\int |u|^t\omega(x)dx\Big)^{1/q} = \Big(\int_B |u(x) - \frac{1}{\omega(B)}\int_B u(y)\omega(y)dy|^t\omega(x)dx\Big)^{1/q}$$

$$< \Big(\int_B \Big(\frac{1}{\omega(B)}\int_B |u(x) - u(y)|\omega(y)dy\Big)^t\omega(x)dx\Big)^{1/q}$$

$$< \Big(\int_B \Big(\frac{1}{\omega(B)}\int_B |u(x) - u(y)|^\alpha\omega(y)dy\Big)^q\omega(x)dx\Big)^{1/q}$$

$$< 2^{\alpha-1}\Big[\Big(\int_B \Big(\frac{1}{\omega(B)}\int_B |u^+(x) - u^+(y)|^\alpha\omega(y)dy\Big)^q\omega(x)dx\Big)^{1/q}$$

$$+ \Big(\int_B \Big(\frac{1}{\omega(B)}\int_B |u^-(x) - u^-(y)|^\alpha\omega(y)dy\Big)^q\omega(x)dx\Big)^{1/q}\Big]$$

$$< 2^{\alpha-1}\Big[\Big(\int_B \Big(\frac{1}{\omega(B)}\int_B |(u^+(x))^\alpha - (u^+(y))^\alpha|\omega(y)dy\Big)^q\omega(x)dx\Big)^{1/q}\Big]$$

$$+ \Big(\int_B \Big(\frac{1}{\omega(B)}\int_B |(u^-(x))^\alpha - (u^-(y))^\alpha|\omega(y)dy\Big)^q\omega(x)dx\Big)^{1/q}\Big]$$

$$< 2^\alpha\Big[\Big(\int_B |(u^+(x))^\alpha - \frac{1}{\omega(B)}\int_B (u^+(y))^\alpha\omega(y)dy|^q\omega(x)dx\Big)^{1/q}$$

$$+ \Big(\int_B |(u^-(x))^\alpha - \frac{1}{\omega(B)}\int_B (u^-(y))^\alpha\omega(y)dy|^q\omega(x)dx\Big)^{1/q}\Big]$$

$$< cR_K^{(\alpha/q)-\beta}R_0^{(d/q)-d+1}\Big(\int_{B\cap\{u>0\}} |\nabla(u^\alpha)|\nu + \int_{B\cap\{u<0\}} |\nabla(u^\alpha)|\nu\Big)$$

$$= cR_K^{(\alpha/q)-\beta}R_0^{(d/q)-d+1} \int_{B(x_0,R_0)} |u|^{\alpha-1}|\nabla u|\nu$$

$$\leq cR_K^{(\alpha/q)-\beta}R_0^{(d/q)-d+1} (\int |u|^{(\alpha-1)p'}\omega)^{1/p'}(\int |\nabla u|^p\sigma)^{1/p} .$$

It is easily seen that $(\alpha - 1)p' = t$, $\frac{1}{q} - \frac{1}{p'} = \frac{1}{t}$, and $\nabla u = \nabla\phi$ so that

$$(\int_B |\phi - \frac{1}{\omega(B)} \int_B \phi\omega|^t\omega)^{1/t} \leq cR_K^{(\alpha/q)-\beta}R_0^{(d/q)-d+1}(\int |\nabla\phi|^p\sigma)^{1/p} .$$

To prove (2.2.61) let $\phi = u^\alpha$ for $\alpha = \frac{t}{q}$ so that

$$(\int |u|^{\alpha q}\omega)^{1/q} \leq \alpha c_2R_K^{(\alpha/q)-\beta}R_0^{(d/q)-d+1}(\int |u|^{\alpha-1}|\nabla u|\nu) ,$$

and so (2.2.61) follows after using Hölder's inequality.

If $K = \bigcup_{i=1}^n M_i$, M_i a compact C^2 manifold of co-dimension $\gamma_i \geq 2$ or a point $(\gamma_i = d)$ then it is routine to show that there exist c_0, r_0, $\varepsilon_0 > 0$ such that

$$H^{d-1}(B(x,r) \cap B(t)) \leq c_0r^{d-\gamma}t^{\gamma-1} \qquad (2.2.67)$$

for $0 < r \leq r_0$, $t \leq \varepsilon_0r$, $\gamma = \text{Min}\ \gamma_i$ and $B(t) = \{x \in \mathbf{R}^d : \text{dist}(x,K) = t\}$. This will be used to verify (2.2.57) and (2.2.58) for all $x \in \mathbf{R}^d$ and $r \in (0,\infty)$ and (2.2.63).

Assume $\frac{2\ \text{diam}\ K}{C} \leq r$ for $C = (\frac{a(d)}{2^{5d+4}a(d-1)})^{1/d-1}$. Let $r' = \text{diam}\ K + \varepsilon_1r$ so if $\varepsilon_1 \leq \frac{c}{2}$, then $A(\varepsilon_1r) \subseteq B(x',r')$ for $x' \in K$ and

$$H^{d-1}(P_z(B(x,r) \cap A(\varepsilon_1r))) \leq H^{d-1}(P_z(A(\varepsilon_1r)))$$

$$\leq H^{d-1}(P_z(B(x',r'))) \leq H^{d-1}(P_z(B(x',cr)))$$

$$\leq a(d-1)(cr)^{d-1} = \frac{a(d)}{2^{5d+4}}r^{d-1}$$

as required in (2.2.57).

If $r \leq r_0$, $t \leq \varepsilon_0r$, and $\sigma > -\gamma$, then the co-area formula implies that

$$\int_{B(x,r)\cap A(t)} \text{dist}^\sigma(x,K)dH^d = \int_0^t s^\sigma H^{d-1}(B(x,r) \cap B(s))ds \qquad (2.2.68)$$

$$\leq c_0r^{d-\gamma} \int_0^t s^{\sigma+\gamma-1}ds \leq \frac{c_0}{\sigma+\gamma} r^{d-\gamma}t^{\sigma+\gamma}$$

2.2.69. By covering K with a finite number of balls of radius less than r_0 and applying (2.2.68) with $\sigma = 0$, it is clear that there exists c', $t_0 > 0$, such that $|A(t)| \le c't^\gamma$ for $t < t_0$.

If $x_1 \in P_z(A(t))$, then choose $x_2 \in A(t)$ such that $P_z(x_2) = x_1$. $P_z^{-1}(x_1) \cap A(2t)$ contains a line segment of length $2t$ since any point within a distance of t from x_2 is in $A(2t)$, so $2tH^{d-1}(P_z(A(t)) \le |A(2t)| \le c'2^\gamma t^\gamma$ and $H^{d-1}(P_z(A(t))) \le c'2^{\gamma-1}t^{\gamma-1}$ for $t < t_0/2$.

If $\frac{r_0}{2} < r < \frac{2 \text{ diam } K}{c}$ and $\varepsilon_1 < \frac{t_0 c}{4 \text{ diam } K}$, then

$$H^{d-1}(P_z(B(x,r) \cap A(r\varepsilon_1))) \le H^{d-1}(P_z(A(r\varepsilon_1)))$$

$$\le c'2^{\gamma-1}\left(\frac{2 \text{ diam } K}{c}\right)^{\gamma-1}\varepsilon_1^{\gamma-1}$$

so that for ε_1 small enough it follows that

$$H^{d-1}(P_z(B(x,r) \cap A(r\varepsilon_1))) \le \frac{a(d)}{2^{6d+3}} r_0^{d-1} \le \frac{a(d)}{2^{5d+4}} r^{d-1}$$

Finally, (2.2.57) will be verified for $r < \frac{r_0}{2}$. If $x_1 \in P_z(B(x,r) \cap A(r\varepsilon_1))$ and $\varepsilon_1 < \min\{1/2, \varepsilon_0/2\}$, then $P_z^{-1}(x_1) \cap B(x,(1 + 2\varepsilon_1)r) \cap A(2r\varepsilon_1)$ contains a line segment of length $2r\varepsilon_1$. Using (2.2.68) with $\sigma = 0$, it now follows that

$$2r\varepsilon_1 H^{d-1}(P_z(B(x,r) \cap A(r\varepsilon_1)))$$

$$\le H^d(B(x,2r) \cap A(2r\varepsilon_1)) \le \frac{c_0}{\gamma} 2^d \varepsilon_1^\gamma r^d$$

and

$$H^{d-1}(P_z(B(x,r) \cap A(r\varepsilon_1))) \le \frac{c_0}{\gamma} 2^{d-1}\varepsilon_1^{\gamma-1} r^{d-1} .$$

It is now clear that (2.2.57) is verified for a sufficiently small choice of ε_1.

Let $\lambda_\sigma(x) = \text{dist}^\sigma(x,K)$ and $\lambda_\sigma(E) = \int_E \lambda_\sigma(x)dx$. It will be shown for $\sigma > -\gamma$ that

$$c^{-1}\lambda_\sigma(B(x,r)) \le \max{}^\sigma\{r, \text{dist}(x,K)\}r^d \le c\lambda_\sigma(B(x,r) - A(\varepsilon_2 r)) \tag{2.2.70}$$

for some c, $\varepsilon_2 > 0$ independent of x and r. (2.2.58) then follows by setting $\sigma = \alpha$. Also, considering $\sigma = p\beta - (p - 1)\alpha$, it follows that if $p\beta - (p-1)\alpha > -\gamma$ in addition to $\alpha > -\gamma$, then

$$R_0\omega^{1/t}(B(x_0,R_0))\sigma^{-1/p}(B(x_0,R_0)) > cR_0R_K^{\alpha/t}R_0^{d/t}R_K^{-\beta+(p-1)\alpha/p}R_0^{-d/p}$$

$$= cR_K^{(\alpha/q)-\beta}R_0^{(d/q)-d+1} ,$$

so (2.2.63) is verified.

The proof of (2.2.70) will be broken down into a number of cases. First consider the case where $r_1 > 2r$ for $r_1 = \text{dist}(x,K)$. An easy calculation shows that

$$\lambda_\sigma < \begin{cases} (r_1 + r)^\sigma & \sigma > 0 \\ (\frac{r_1}{2})^\sigma & \sigma < 0 \end{cases}$$

on $B(x,r)$, so $\lambda_\sigma(B(x,r)) < c \max^\sigma\{r_1,r\}r^d$. Similarly it is seen that $\lambda_\sigma(B(x,r)) > c \max^\sigma\{r_1,r\}r^d$. If $\varepsilon_2 < 1$, then $B(x,r) \cap A(\varepsilon_2 r) = \emptyset$ so $\lambda_\sigma(B(x,r) - A(\varepsilon_2 r)) > c \max^\sigma\{r_1,r\}r^d$.

If $r_1 < 2r$ and $r < r_0$, then it follows from (2.2.68) that

$$\lambda_\sigma(B(x,r) \cap A(\varepsilon_0 r)) < \frac{c_0 r^{d+\sigma}\varepsilon_0^{\gamma+\sigma}}{\gamma + \sigma} .$$

Also,

$$\lambda_\sigma < \begin{cases} (3r)^\sigma & \sigma > 0 \\ (\varepsilon_0 r)^\sigma & \sigma < 0 \end{cases}$$

on $B(x,r) - A(\varepsilon_0 r)$, so

$$\lambda_\sigma(B(x,r)) < \lambda_\sigma(B(x,r) - A(\varepsilon_0 r)) + \lambda_\sigma(B(x,r) \cap A(\varepsilon_0 r))$$

$$< cr^{\sigma+d} < c \max^\sigma\{r_1,r\}r^d .$$

If $\varepsilon_2 < \varepsilon_0$, then $|B(x,r) \cap A(\varepsilon_2 r)| < c\varepsilon_2^\gamma r^d$ so if ε_2 is small enough, then $|B(x,r) - A(\varepsilon_2 r)| > cr^d$. Now considering the two cases $\sigma < 0$ and $\sigma > 0$ separately, it follows that

$$\lambda_\sigma(B(x,r) - A(\varepsilon_2 r)) > cr^{d+\sigma} > c \max^\sigma\{r_1,r\}r^d .$$

Covering K with balls as in 2.2.69 and using (2.2.68) it follows that there exists $t_0 > 0$ such that $\lambda_\sigma(A(t)) < ct^{\sigma+\gamma}$ for $t < t_0$. If $r_0 < r < 2 \text{ diam } K$, then

$$\lambda_\sigma(B(x,r)) \le \lambda_\sigma\left(B(x,r) - A\left(\frac{t_0 r}{2 \text{ diam } K}\right)\right) + \lambda_\sigma\left(A\left(\frac{t_0 r}{2 \text{ diam } K}\right)\right)$$

$$\le c(r^{d+\sigma} + t_0^{\sigma+\gamma}) \le c\left(1 + \frac{t_0^\gamma}{r_0^{d+\sigma}}\right)r^{d+\sigma} \le c \max^\sigma\{r_1, r\}r^d \ .$$

As in 2.2.69, $|A(t)| \le ct^\gamma$ for $t \le t_0$, so if $\varepsilon_2 \le t_0(2 \text{ diam } K)^{-1}$, then

$$|B(x,r) - A(\varepsilon_2 r)| \ge |B(x,r)| - |A(\varepsilon_2 r)|$$

$$\ge a(d)r_0^d - c\varepsilon_2^\gamma(2 \text{ diam } K)^\gamma \ ,$$

and so for ε_2 small enough

$$|B(x,r) - A(\varepsilon_2 r)| \ge c(2 \text{ diam } K)^d \ge cr^d \ ,$$

and

$$\lambda_\sigma(B(x,r) - A(\varepsilon_2 r)) \ge cr^{\sigma+d} \ge c \max^\sigma\{r_1, r\}r^d \ .$$

Finally, if $r > 2 \text{ diam } K$, then for some $x' \in K$ use polar coordinates to get

$$\lambda_\sigma(B(x,r)) \le \lambda_\sigma(B(x', 2 \text{ diam } K)) + \lambda_\sigma(B(x', 4r) - B(x', 2 \text{ diam } K))$$

$$\le c\left((2 \text{ diam } K)^{d+\sigma} + \int_{2 \text{ diam } K}^{4r} s^{\sigma+d-1}ds\right)$$

$$\le cr^{d+\sigma} \le c \max^\sigma\{r_1, r\}r^d \ .$$

Also, $A(\varepsilon_2 r) \subseteq B(x', r')$ for $x' \in K$ and $r' = \text{diam } K + \varepsilon_2 r$, so if $\varepsilon_2 \le \frac{1}{4}$, then $r' \le \frac{3}{4} r$, and $|B(x,r) - A(\varepsilon_2 r)| \ge a(d)(1 - (\frac{3}{4})^d)r^d$, so

$$\lambda_\sigma(B(x,r) - A(\varepsilon_2 r)) \ge c \max^\sigma\{r_1, r\}r^d \ . \quad \blacksquare$$

The Euler Equation for Capacitary Extremals and a Wirtinger Inequality

The C_H-capacitary extremal for $E \subseteq \Omega$ satisfies a degenerate differential equation. This can be used to develop an interesting sufficient condition for a special case of inequality (2.2.46). This is motivated by a paper of Meyers [MY1].

If $(u, \nabla u) \in W_0^{1,p}(\Omega)$ and μ is a finite positive Borel measure, then it is said that $(u, \nabla u)$ satisfies $-\text{div}(\nu|\nabla u|^{p-2}\nabla u) = \mu$ weakly if $\int \nabla\phi \cdot \nabla u |\nabla u|^{p-2} d\nu = \int \phi d\mu$ for all $\phi \in C_0^\infty(\Omega)$. The convention $0 \cdot \infty = 0$ is used for $|\nabla u|^{p-2}\nabla u$ if $1 \le p < 2$.

2.2.71 Proposition. Assume $1 < p < \infty$ and inequality (2.2.42) holds with $p = q$. If $E \subset \Omega$, $\bar{E}$ compact, and $0 < C_H(E) < \infty$ for $H = C_0^\infty(\Omega)$, then there exists a finite positive Borel measure μ supported on ∂E such that $\mu(\Omega) = C_H(E)$ and $\mu(F) = 0$ if $C_H(F) = 0$ and F is Borel measurable. Also,

$$|\int \sigma d\mu| \leq c(\int |\nabla\sigma|^p d\nu + \int |\sigma|^p d\nu)^{1/p}$$

for all $\sigma \in C^\infty(\Omega)$ and $c = C_H^{(p-1)/p}(E)\max\{\text{dist}^{-1}(E,\partial\Omega),1\}$, and

$$-\text{div}(\nu|\nabla u|^{p-2}\nabla u) = \mu \quad \text{weakly}$$

if $(u,\nabla u)$ is the capacitary extremal of E.

Remarks. A weaker notion of Sobolev space can be developed which admits much of Proposition 2.2.71 without assuming the Poincaré inequality (2.2.42).

The dependence of c on $\text{dist}(E,\partial\Omega)$ can be removed in the nonweighted case if an alternate capacity is used.

Using Lagrange multiplier techniques it can be shown that $\bar{C}_{\bar{H}}$ capacitary extremals of compact sets satisfy a similar differential equation.

Proof. Let $(u,\nabla u)$ be the capacitary extremal of E which exists by (2.2.33) and (2.2.34), and let $F(t,\sigma) = \int |\nabla u + t\sigma|^p d\nu$ for $t \in \mathbf{R}$ and $\sigma \in C_0^\infty(\Omega)$. By the mean value theorem if $x,y \in \mathbf{R}^n$, then

$$\frac{|x + ty|^p - |x|^p}{t} = p|x + t^*y|^{p-2}(x + t^*y)\cdot y$$

for some t^* between 0, t, where the convention $0\cdot\infty = 0$ is used for $|x + t^*y|^{p-2}(x + t^*y)$ if $1 < p < 2$ and $x + t^*y = 0$. It is clear then that

$$\frac{|\nabla u + t\nabla\sigma|^p - |\nabla u|^p}{t} \to p|\nabla u|^{p-2}\nabla u\cdot\nabla\sigma$$

ν almost everywhere, the difference quotient being dominated by $p(|\nabla u| + t|\nabla\sigma|)^{p-1}|\nabla\sigma|$, which is seen to be in $L^p(\nu,\Omega)$ by using Hölder's inequality and recalling that $|\nabla\sigma|$ has compact support and ν is locally finite. From the dominated convergence theorem it now follows that

$$\frac{d}{dt}\bigg|_{t=0} = \lim_{t\to 0}\frac{F(t,\sigma) - F(0,\sigma)}{t} = p\int |\nabla u|^{p-2}\nabla u\cdot\nabla\sigma\, d\nu\,. \tag{2.2.72}$$

2.2.73. If $u + t\sigma \geq 1$ quasi-everywhere on E, then by Proposition 2.2.30 it follows that

$$F(0,\sigma) = \int |\nabla u|^p d\nu = C_H(E) \le \int |\nabla u + t\nabla\sigma|^p d\nu = F(t,\sigma) .$$

If $\sigma \geq 0$ and $t \geq 0$, then $u + t\sigma \geq 1$ quasi-everywhere on E and

$$0 \le \left.\frac{dF}{dt}\right|_{t=0} = p \int |\nabla u|^{p-2}\nabla u\cdot\nabla\sigma \, d\nu . \tag{2.2.74}$$

Let $T(\sigma) = \int |\nabla u|^{p-2}\nabla u\cdot\nabla\sigma \, d\nu$ for $\sigma \in C_0^\infty(\Omega)$. By Hölder's inequality it follows that $|T(\sigma)| \le C_H^{(p-1)/p}(E)(\int |\nabla\sigma|^p d\nu)^{1/p}$, so that T is a distribution. From (2.2.74) it follows that T is positive, and so T is a locally finite positive Borel measure μ on Ω, that is, $\int |\nabla u|^{p-2}\nabla u\cdot\nabla\sigma \, d\nu = \int \sigma d\mu$ and so $-\mathrm{div}(\nu|\nabla u|^{p-2}\nabla u) = \mu$ weakly. If the support of $\sigma \subseteq \Omega - \bar{E}$, then for any $t \in \mathbf{R}$, $u + t\sigma \geq 1$ quasi-everywhere on E and so $T(\sigma) = 0$ by (2.2.72) and 2.2.73, but then μ is supported in $\bar{E}$ and so is finite since $\bar{E}$ is compact.

If $K \subseteq \Omega$ is compact and $C_H(K) = 0$, then choose $\phi_n \in C_0^\infty(\Omega)$ such that $\phi_n \geq 1$ on K and $\int |\nabla\phi_n|^p d\nu \to 0$ so

$$\mu(K) \le \int \phi_n d\mu = \int |\nabla u^{p-2}\nabla u\cdot\nabla\phi_n d\nu$$

$$\le (\int |\nabla u|^p d\nu)^{(p-1)/p}(\int |\nabla\phi_n|^p d\nu)^{1/p} \to 0 .$$

If $F \subseteq \Omega$ is Borel measurable and $C_H(F) = 0$, use the regularity of μ to choose $K_n \subseteq F$, K_n compact such that $\mu(K_n) \to \mu(F)$. $C_H(K_n) \le C_H(F) = 0$ so $\mu(K_n) = 0$ as above, and therefore $\mu(F) = 0$.

Choose $\phi_n \in C_0^\infty(\Omega)$ such that $\phi_n \to u$ in $W_0^{1,p}(\Omega)$. As in the proof of Proposition 2.2.7, the ϕ_n can be chosen to be uniformly bounded and by (2.2.23) they can be chosen to converge pointwise quasi-everywhere to u. Therefore

$$\int u d\mu = \lim_{n\to\infty} \int \phi_n d\mu = \lim_{n\to\infty} \int |\nabla u|^{p-2}\nabla u\cdot\nabla\phi_n d\nu = \int |\nabla u|^p d\nu = C_H(E) .$$

If E is compact, then since $u = 1$ quasi-everywhere on E and μ is supported on E, it follows that $\mu(\Omega) = \mu(E) = \int u d\mu = C_H(E)$. Otherwise use Proposition 2.2.15 to choose $K_n \subseteq E$ such that $K_n \subseteq K_{n+1}$ and $C_H(K_n) \to C_H(E)$. If u_n is the capacitary extremal for K_n and μ_n the associated measure, then an application of Clarkson's inequalities as in the proof of (2.2.30) implies that the u_n are Cauchy in $W_0^{1,p}(\Omega)$. By (2.2.32) there exists a $\bar{u} \in W_0^{1,p}(\Omega)$, $\bar{u}$ quasicontinuous and a subsequence $\{n_i\}$ such that $u_{n_i} \to \bar{u}$ in $W_0^{1,p}(\Omega)$ and pointwise quasi-everywhere. $C_H(E) = \lim_{n\to\infty} C_H(K_n) \le$

$C_H(\bigcup K_n) \leq C_H(E) = \int |\nabla\bar{u}|^p d\nu$ and $\bar{u} = 1$ quasi-everywhere on $\bigcup K_n$, since $K_n \subseteq K_{n+1}$, so $\bar{u}$ is a capacitary extremal for $\bigcup K_n$ but u is as well, for similar reasons so $u = \bar{u}$ in $W_0^{1,p}(\Omega)$ by (2.2.34), and so $u_n \to u$. Choose $\sigma \in C_0^\infty(\Omega)$ with $\sigma = 1$ on $\bar{E}$ so that

$$C_H(E) = \lim_{n\to\infty} C_H(K_n) = \lim_{n\to\infty} \mu_n(K_n)$$

$$= \lim_{n\to\infty} \int \sigma d\mu_n = \lim_{n\to\infty} \int |\nabla u_n|^{p-2}\nabla u_n \cdot \nabla\sigma$$

$$= \int |\nabla u|^{p-2}\nabla u \cdot \nabla\sigma \, d\nu = \int \sigma d\mu = \mu(\bar{E}) .$$

To complete the proof that μ is supported in ∂E consider $h_{-\infty,1}$ as in 2.2.5 so that $(h_{-\infty,1}(u), \chi_{\{u<1\}}\nabla u) \in W_0^{1,p}(\Omega)$. It is clear that this is quasicontinuous by inspection of the proof of 2.2.5. Also $h_{-\infty,1}(u) = 1$ quasi-everywhere on E, so

$$C_H(E) \leq \int |\chi_{\{u<1\}}\nabla u|^p d\nu \leq \int |\nabla u|^p d\nu = C_H(E) ,$$

therefore $\nabla u = 0$, ν almost everywhere on $\{u \geq 1\}$. Since u may be chosen to be one everywhere on E by (2.2.33), it is seen that $\nabla u = 0$, ν almost everywhere on E.

Choose $\sigma_n \in C_0^\infty(\Omega)$, $n = 0,1,2,\ldots$ such that $0 \leq \sigma_n \leq 1$, $\sigma_0 = 1$ on $\bar{E}$, $\sigma_n(x) = \sigma_0(x)$ for $x \in \Omega$-interior E and $\lim_{n\to\infty} \sigma_n(x) = 0$ for $x \in$ interior E.

$$\mu(\partial E) = \lim_{n\to\infty} \int \sigma_n d\mu = \lim_{n\to\infty} \int |\nabla u|^{p-2}\nabla u \cdot \nabla\sigma_n d\nu$$

$$= \lim_{n\to\infty} \int |\nabla u|^{p-2}\nabla u \cdot \nabla\sigma_0 d\nu \quad \text{since } \nabla u = 0, \ \nu \text{ almost everywhere on interior of } E$$

$$= \int \sigma d\mu = \mu(\bar{E}).$$

Finally, given $\sigma \in C^\infty(\Omega)$, choose ϕ such that $\phi \in C_0^\infty(\Omega)$, $0 \leq \phi \leq 1$, and $\phi = 1$ on $\bar{E}$.

$$|\int \sigma d\mu| = |\int \sigma\phi \, d\mu| = |\int |\nabla u|^{p-2}\nabla u \cdot \nabla(\sigma\phi) d\nu|$$

$$\leq \left(\int |\nabla u|^p d\nu\right)^{(p-1)/p}\left(\int (|\nabla\sigma|^p \phi^p + |\nabla\phi|^p |\sigma|^p) d\nu\right)^{1/p}$$

$$\leq \max\{1, \sup|\nabla\phi|\} C_H^{(p-1)/p}(E)\left(\int (|\nabla\sigma|^p + |\sigma|^p) d\nu\right)^{1/p}$$

and a $C_0^\infty(\Omega)$ smoothing of $1 - \frac{\text{dist}(x,E)}{\text{dist}(E,\partial\Omega)}$ along with a limiting process gives the necessary constant. ∎

2.2.75 Proposition. Assume (2.2.42) for some $q > p > 1$, $\lambda = \omega = \nu$, $\lambda(\Omega) = 1$. If T is a linear functional on $C^\infty(\Omega) \cap W^{1,p}(\Omega)$ such that $|T\phi| \leq c_1(\int |\nabla\phi|^p d\nu + \int |\phi|^p d\nu)^{1/p}$ and $T(1) = 1$, then

$$\left(\int |\phi - T(\phi)|^q d\nu\right)^{1/q} \leq c_2\left(\int |\nabla\phi|^p d\nu\right)^{1/p} .$$

Proof. If $\phi \in C^\infty(\Omega)$, then

$$\begin{aligned}
&\left(\int |\phi - T(\phi)|^q d\nu\right)^{1/q} \\
&\quad \leq \left(\int|\phi - \int \phi d\nu|^q d\nu\right)^{1/q} + \left(\int |T(\phi - \int \phi d\nu)|^q d\nu\right)^{1/q} \\
&\quad \leq c\left(\int |\nabla\phi|^p d\nu\right)^{1/p} + |T(\phi - \int \phi d\nu)| \quad \text{since } \nu(\Omega) = 1 \\
&\quad \leq c\left(\int |\nabla\phi|^p d\nu\right)^{1/p} + c_1\left(\int (|\nabla\phi|^p + |\phi - \int \phi d\nu|^p) d\nu\right)^{1/p} \\
&\quad \leq c\left(\int |\nabla\phi|^p d\nu\right)^{1/p} . \quad \blacksquare
\end{aligned}$$

2.2.76 Proposition. Assume $1 < p < \infty$, $\lambda = \omega = \nu$, $\lambda(\Omega) = 1$, inequality (2.2.46) holds for $q = r$, and inequality (2.2.42) holds for $q = p$. If $E \subseteq \Omega$, $\bar{E}$ compact and $0 < C_H(\bar{E}) < \infty$, then

$$\left(\int |\phi - \int \phi d\mu|^r d\nu\right)^{1/r} \leq c_1\left(\int |\nabla\phi|^p d\nu\right)^{1/p}$$

for all $\phi \in C^\infty(\Omega)$, where μ is the measure associated with E as in Proposition 2.2.71 and $c_1 = c(1 + \max\{\operatorname{dist}^{-1}(E,\partial\Omega),1\}C_H^{-1/p}(E))$.

Consequently if $\phi \in C^\infty(\Omega)$ and $\phi = 0$ on E, then

$$\left(\int |\phi|^q d\nu\right)^{1/q} \leq c_1\left(\int |\nabla\phi|^p d\nu\right)^{1/p} .$$

Proof. Let $T(\phi) = C_H^{-1}(E) \int \phi d\mu$ and use Propositions 2.2.71 and 2.2.75. ■

Boundary Values for $W_0^{1,p}$ Functions

The following proposition is a generalization to weighted spaces of a result of Bagby [BG].

2.2.77 Proposition. Suppose $\Omega' \subseteq R^d$ is open and bounded, $\Omega' \subseteq \Omega$, and

$$\int |\phi|^p d\omega \leq c \int |\nabla\phi|^p d\nu \tag{2.2.78}$$

for all $\phi \in C_0^\infty(\Omega)$.

If $(u,\nabla u) \in W_0^{1,p}(\Omega)$, u is quasicontinuous and $u = 0$ quasi-everywhere in $\Omega - \Omega'$, then

$$(u,\chi_{\{u\neq 0\}}\nabla u)|_{\Omega'} \in W_0^{1,p}(\Omega')$$

and is quasicontinuous with respect to Ω'. If in addition $\nabla u = 0$, ν almost everywhere on $\{u = 0\}$, then

$$(u,\nabla u)|_{\Omega'} \in W_0^{1,p}(\Omega') .$$

Proof. At first it is assumed that $|u| < M < \infty$. Inequality (2.2.78) implies as in Proposition 2.2.41 that $\omega(E) < cC_H(E)$ for Borel sets $E \subseteq \Omega$ and $H = C_0^\infty(\Omega)$, so a Borel set of zero capacity always has zero ω measure.

2.2.79 It is clear then that u may be altered on a set of C_H capacity zero so that $u = 0$ everywhere on $\Omega - \Omega'$. By Definition 2.2.20 u is still quasicontinuous and there exist $\bar{\phi}_n \in C_0^\infty(\Omega)$ such that $\bar{\phi}_n \to u$ in $W_0^{1,p}(\Omega)$ and pointwise quasi-everywhere. As in the proof of 2.2.7 the $\bar{\phi}_n$ can be chosen so that $|\bar{\phi}_n| < M$.

2.2.80. By 2.2.5 and 2.2.4 there exist $f_n \in C^\infty(\mathbf{R})$ with $0 < f_n(x) < x^+$ and a subsequence n_m such that if $\phi_m = f_m(\bar{\phi}_{n_m})$, then $(\phi_m,\nabla\phi_m) \to (u^+,\chi_{\{u>0\}}\nabla u)$ in $W_0^{1,p}(\Omega)$. Since $\bar{\phi}_{n_m} \to u$ quasi-everywhere, it follows from 2.2.4 that $\phi_n \to u^+$ quasi-everywhere.

2.2.81. Therefore $(u^+,\chi_{\{u>0\}})$ is quasicontinuous (with respect to Ω). The ϕ_n will be used to construct a sequence $\{u_n\}$ in $W_0^{1,p}(\Omega)$ such that $u_n < 0$ on $\Omega' - c_n$, where $c_n \subseteq \Omega'$ is compact, and $(u_n ,\nabla u_n) \to (u^+,\chi_{\{u>0\}}\nabla u)$ in $W_0^{1,p}(\Omega)$. Using this it will be shown that $(u^+,\chi_{\{u>0\}}\nabla u)|_{\Omega'} \in W_0^{1,p}(\Omega')$.

Choose a bounded open set $\Omega'' \subseteq \mathbf{R}^d$ such that $\bar{\Omega}' \subseteq \Omega''$ and $\bar{\Omega}'' \subseteq \Omega$. In addition choose $\rho \in C_0^\infty(\Omega)$ such that $0 < \rho < 1$ and $\rho = 1$ on Ω''.

2.2.82. Given $\varepsilon > 0$, pick m s.t. $\|\phi_n - u^+\|_{1,p;\Omega} < \varepsilon$ for all $n > m$, and choose N, N' open such that $\partial\Omega' \subseteq N$, $\bar{N} \subseteq N'$, $\bar{N} \subseteq \Omega''$, and

$$\int_{N'\cap\{u>0\}} |\nabla u|^p d\nu < \varepsilon . \tag{2.2.83}$$

This can be done since $u = 0$ on $\Omega - \Omega'$ and so in particular on $\partial\Omega'$.

2.2.84. Choose $\alpha > 0$ such that $\|\alpha\rho\|_{1,2;\Omega} = \alpha\|\rho\|_{1,2;\Omega} < \varepsilon$, and choose $\sigma \in C_0^\infty(N')$ such that $\sigma = 1$ on N. The ϕ_n converge ν almost everywhere to u^+ since ν is absolutely continuous with respect to ω so arguing as in the proof of 2.2.7 it follows that $\sigma\phi_n \to \sigma u^+$ in $W_0^{1,p}(N')$ and so by 2.2.19 if $H' = C_0^\infty(N')$, then for some subsequence $\{n_i\}$, the $\sigma\phi_{n_i}$ converge uniformly off open sets of arbitrarily small $C_{H'}$ capacity. Let $E = \{x \in N' : \phi_{n_i}(x) \not\to u^+(x)\}$. Since u^+ is quasicontinuous with respect to Ω, it follows that $C_H(E) = 0$ for $H = C_0^\infty(\Omega)$ and (2.2.35) implies that $C_{H'}(E) = 0$. Since E is contained in open subsets of N' with arbitrarily small $C_{H'}$ capacity, it follows that the $\sigma\phi_{n_i}$ converge uniformly to σu^+ off open subsets of N' of arbitrarily small $C_{H'}$-capacity. Let G be an open subset of N' such that $C_{H'}^{1/p}(G) < \varepsilon$ and $|\sigma\phi_{n_i} - \sigma u^+| < \frac{\alpha}{2}$ in $N' - G$ for $i > i_\alpha$. On $\partial\Omega$, $u^+ = 0$ and $\rho = \sigma = 1$, so $\phi_{n_i} - \alpha\rho < -\frac{\alpha}{2}$ on $\partial\Omega - G$, in which case $\phi_{n_i} - \alpha\rho < 0$ on an open set H_{n_i} with $\partial\Omega - G \subseteq H_{n_i}$. Let W be a capacitary extremal for G relative to $C_0(N')$ as in (2.2.33) such that

$$0 \le W \le 1,\ W = 1 \text{ on } G, \text{ and } \int |\nabla W|^p dv = C_{H'}(G) < \varepsilon^p . \quad (2.2.85)$$

Let $V = (\phi_{n_1} - \alpha\rho)(1 - W)$, where $i > i_\alpha$ and $n_i > m$ so $V < 0$ on $H_{n_i} \cup G$ and $V \in W_0^{1,p}(\Omega)$ by 2.2.9. Using (2.2.78), 2.2.82, (2.2.83), 2.2.84, and (2.2.85) it follows that

$$\begin{aligned}
\|V - u^+\|_{1,p;\Omega} &\le \|\phi_{n_i} - u\|_{1,p;\Omega} + \|\alpha\rho\|_{1,p;\Omega} + \|(\phi_{n_i} - \alpha\rho)W\|_{1,p;\Omega} \\
&\le 2\varepsilon + (c+1)^{1/p}\left(\int |\nabla[(\phi_{n_i} - \alpha\rho)W]|^p dv\right)^{1/p} \\
&\le 2\varepsilon + (c+1)^{1/p}\left(M + \frac{\varepsilon}{\|\rho\|_{1,p;\Omega}}\right)\left(\int |\nabla w|^p dv\right)^{1/p} \\
&\quad + (c+1)^{1/p}\left(\int_{N'} |\nabla\phi_{n_i}|^p dv\right)^{1/p} \\
&\le \left[2 + (c+1)^{1/p}\left(M + \frac{\varepsilon}{\|\rho\|_{1,p;\Omega}}\right)\right]\varepsilon \\
&\quad + (c+1)^{1/p}\left[\left(\int_{N'\cap\{u>0\}} |\nabla u|^p dv\right)^{1/p}\right. \\
&\quad \left. + \left(\int |\nabla\phi_{n_i} - \chi_{\{u>0\}}\nabla u|^p dv\right)^{1/p}\right] \\
&\le \left(2 + (c+1)^{1/p}\left[2 + M + \frac{\varepsilon}{\|\rho\|_{1,2;\Omega}}\right]\right)\varepsilon .
\end{aligned}$$

Since ε is arbitrary and $\Omega' - (H_{n_i} \cup G)$ is a compact subset of Ω', $(\partial\Omega' \subseteq H_{n_i} \cup G)$, it is clear that there exist $u_n \in W_0^{1,p}(\Omega)$ such that u_n is bounded, $u_n \to u^+$ in $W_0^{1,p}(\Omega)$, and $u_n \le 0$ on $\Omega' - c_n$ for some compact subset c_n of Ω'.

Let f_n be as in 2.2.80 so that for some subsequence $\{n_m\}$, $(f_m(u_{n_m}), f_m'(u_{n_m})\nabla u_{n_m}) \to (u^+, \chi_{\{u>0\}}\nabla u)$ in $W_0^{1,p}(\Omega)$. It is claimed that $(f_m(u_{n_m}), f_m'(u_{n_m})\nabla u_{n_m})\big|_{\Omega'} \in W_0^{1,p}(\Omega')$ so it follows that $(u^+, \chi_{\{u>0\}}\nabla u)\big|_{\Omega'} \in W_0^{1,p}(\Omega')$. The same procedure applied to $-u$ shows that $(-(-u)^+, \chi_{\{u<0\}}\nabla u)\big|_{\Omega'} \in W_0^{1,p}(\Omega')$ so that addition gives $(u, \chi_{\{u\neq 0\}}\nabla u)\big|_{\Omega'} \in W_0^{1,p}(\Omega')$. To prove the claim choose $\eta_m \in C_0^\infty(\Omega')$ such that $\eta_m = 1$ on c_{n_m}. By 2.2.9

$$(\eta_m f_m(u_{n_m}), f_m(u_{n_m})\nabla\eta + \eta_m f_m'(u_{n_m})\nabla u_{n_m})\big|_{\Omega'} \in W_0^{1,p}(\Omega') ,$$

but $f_m(x) = 0$ if $x \le 0$, so $f_m(u_{n_m}) = 0$ on $\Omega' - c_{n_m}$ and also $f_m'(u_{n_m}) = 0$ on $\Omega' - c_{n_m}$ for the same reason, so this, combined with the fact that $\eta_m = 1$ and $\nabla\eta_m = 0$ on c_{n_m}, implies that $(f_m(u_{n_m}), f_m'(u_{n_m})\nabla u_{n_m}\big|_{\Omega'} \in W_0^{1,p}(\Omega')$, as claimed.

Considering 2.2.81 and (2.2.35) it is clear that $(u, \chi_{\{u\neq 0\}})\big|_{\Omega'}$ is quasicontinuous with respect to Ω'. If u is unbounded, then let $f_n = h_{-N,N}$ as in 2.2.5 so that $(f_N(u), \chi_{\{|u|<N\}}\nabla u) \in W_0^{1,p}(\Omega)$ and is quasicontinuous because of 2.2.4 and 2.2.5. Also $f_N(u)$ is bounded and zero quasi-everywhere in $\Omega - \Omega'$ so that the work above implies that $(f_N(u), \chi_{\{|u|<N, u\neq 0\}}\nabla u)\big|_{\Omega'} \in W_0^{1,p}(\Omega')$, and so $(u, \chi_{\{u\neq 0\}}\nabla u)\big|_{\Omega'} \in W_0^{1,p}(\Omega')$ since $\int |f_N(u) - u|^p d\omega = \int_{\{|u|>N\}} |u|^p d\omega \to 0$ as $N \to \infty$ and $\int |\chi_{\{|u|<N, u\neq 0\}}\nabla u - \chi_{\{u\neq 0\}}\nabla u|^p d\nu = \int_{|u|>N} |\nabla u|^p d\nu \to 0$ as $N \to \infty$. Since $(f_N(u), \chi_{\{|u|<N\}}\nabla u)\big|_{\Omega'}$ is also quasicontinuous with respect to Ω' and $f_N(u) \to u$ everywhere, it follows from (2.2.32) that $(u, \chi_{\{u\neq 0\}}\nabla u)$ is quasicontinuous with respect to Ω'. ∎

<u>Weak Boundary Values</u>

If $f : \partial\Omega \to \mathbf{R}$ is continuous, $u \in W^{1,p}(\Omega)$ and $x \in \partial\Omega$, then it is said that $u(x) \le \ell$ weakly if for every $k > \ell$ there exists an $r > 0$ such that $(\eta(u-k)^+, \eta\chi_{\{u>k\}}\nabla u + (u-k)^+\nabla\eta) \in W_0^{1,p}(\Omega')$ for all $\eta \in C_0^\infty(B(x,r))$. In addition $u(x) \ge \ell$ weakly if $-u(x) \le -\ell$ weakly and $u(x) = \ell$ weakly if both $u(x) \le \ell$ weakly and $u(x) \ge \ell$ weakly. Proposition 2.2.86 shows that under certain conditions this definition of weak boundary values is virtually equivalent to a more conventional definition.

2.2.86 Proposition. Suppose Ω' is open and bounded, $\bar{\Omega}' \subseteq \Omega$, $u \in W^{1,p}(\Omega')$, $f \in W_0^{1,p}(\Omega)$, f quasicontinuous in Ω and continuous in $\Omega - \Omega'$, and either $\nu < c\omega$ or u, f are bounded.

If $(u - f|_{\Omega'}, \nabla u - \nabla f|_{\Omega'}) \in W_0^{1,p}(\Omega')$, then $u(x) = f(x)$ weakly for all $x \in \partial\Omega'$. Conversely if $u(x) = f(x)$ weakly for all $x \in \partial\Omega'$, then $(u - f|_{\Omega'}, \chi_{\{u \neq f\}}(\nabla u - \nabla f|_{\Omega'})) \in W_0^{1,p}(\Omega')$.

Remark. In the converse it is only necessary to assume that $u(x) = f(x)$ weakly quasi-everywhere on $\partial\Omega'$. This follows from an argument similar to that in Proposition 2.2.77, where a capacitary extremal is used to remove an open set of small capacity.

Proof. Assume $(u - f|_{\Omega'}, \nabla u - \nabla f|_{\Omega'}) \in W_0^{1,p}(\Omega')$. Choose $\phi_n \in C_0^\infty(\Omega')$ such that $\phi_n \to u - f|_{\Omega'}$ in $W_0^{1,p}(\Omega')$. Consider the ϕ_n as functions in $C_0^\infty(\Omega)$, let v be a quasicontinuous limit of the ϕ_n in $W_0^{1,p}(\Omega)$ and $\bar{u} = v + f$, so $\bar{u} = f$ quasi-everywhere on $\Omega - \Omega'$. If $\psi_n \in C_0^\infty(\Omega)$ such that $\psi_n \to f$ in $W_0^{1,p}(\Omega)$, then $(\phi_n + \psi_n, \nabla\phi_n + \nabla\psi_n) \to (\bar{u}, \nabla\bar{u})$ in $W_0^{1,p}(\Omega)$ with $\nabla\bar{u} = \nabla v + \nabla f$ so $(\phi_n + \psi_n|_{\Omega'}, \nabla\phi_n + \nabla\psi_n|_{\Omega'}) \to (\bar{u}, \nabla\bar{u}|_{\Omega'})$ in $W^{1,p}(\Omega')$, but $(\phi_n + \psi_n|_{\Omega'}, \nabla\phi_n + \nabla\psi_n|_{\Omega'}) \to (u, \nabla u)$ in $W^{1,p}(\Omega')$ also, so

$$\begin{cases} \bar{u} = u, & \omega \text{ almost everywhere, and} \\ \nabla\bar{u} = \nabla u, & \nu \text{ almost everywhere .} \end{cases} \tag{2.2.87}$$

Given $x_0 \in \partial\Omega'$, pick K so that $f(x_0) < K$. Since f is continuous in $\Omega - \Omega'$, there eixsts an $r > 0$ such that $f(x) < K$ in $B(x_0, r) \cap (\Omega - \Omega')$, so if $\eta \in C_0^\infty(B(x_0, r))$, then $\eta(\bar{u} - K)^+ = \eta(f - K)^+ = 0$ quasi-everywhere in $\Omega - \Omega'$.

2.2.88. $\bar{u} - K \in W_{loc}^{1,p}(\Omega)$ so that 2.2.5 implies that

$$((\bar{u} - K)^+, \chi_{\{\bar{u}>K\}}\nabla\bar{u}) \in W_{loc}^{1,p}(\Omega) ,$$

and so 2.2.9 implies that

$$(\eta(\bar{u} - K)^+, \eta\chi_{\{\bar{u}>K\}}\nabla\bar{u} + (\bar{u} - K)^+\nabla\eta) \in W_0^{1,p}(\Omega) .$$

Following the proofs of 2.2.5 and 2.2.9, it is clear that $(\bar{u} - K)^+$ is locally quasicontinuous (2.2.20) and $\eta(\bar{u} - K)^+$ is quasi-continuous. Proposition 2.2.77 now implies that

$$(\eta(\bar{u} - K)^+, \eta\chi_{\{\bar{u}>0\}}\nabla\bar{u} + (\bar{u} - K)^+\nabla\eta)|_{\Omega'} \in W_0^{1,p}(\Omega') ,$$

but (2.2.87) then implies that

$$(\eta(u - K)^+, \eta\chi_{\{u>0\}}\nabla u + (u - K)^+\nabla\eta)|_{\Omega'} \in W_0^{1,p}(\Omega')$$

since ν is absolutely continuous to ω. This is true for all $K > f(x_0)$ so $u(x_0 \leq f(x_0)$ weakly. In the same manner it is shown that $u(x_0 \geq f(x_0)$ weakly and so $u(x_0) = f(x_0)$ weakly for all $x_0 \in \partial\Omega'$.

Conversely if $u(x) = f(x)$ weakly for all $x \in \partial\Omega$, then for $x_0 \in \partial\Omega'$ and $\varepsilon > 0$, there exists an $r > 0$ such that $|f(x) - f(x_0)| < \varepsilon$ for $x \in \mathbf{B}(x_0,r) \cap (\Omega - \Omega')$ and $\eta(u - f(x_0) - \varepsilon)^+ \in W_0^{1,p}(\omega)$ for all $\eta \in C_0^\infty(\mathbf{B}(x_0,r))$. From the first inequality it follows that $-f(x) + f(x_0) - \varepsilon < 0$ on $\mathbf{B}(x_0,r) \cap (\Omega - \Omega')$ so $\eta(-f + f(x_0) - \varepsilon)^+ = 0$ on $\Omega - \Omega'$ for $\eta \in C_0^\infty(\mathbf{B}(x_0,r))$. As in 2.2.88, it follows that $\eta(-f + f(x_0) - \varepsilon)^+|_{\Omega'} \in W_0^{1,p}(\Omega')$. $\partial\Omega'$ is compact since Ω' is bounded so a covering of balls such as $\mathbf{B}(x_0,r/2)$ can be reduced to a finite subcover $\mathbf{B}(x_i,r_i/2)$, $i = 1,\ldots,n$, such that $\eta(u - f(x_i) - \varepsilon)^+$ and $\eta(-f + f(x_i) - \varepsilon)^+$ are in $W_0^{1,p}(\Omega')$ for $\eta \in C_0^\infty(\mathbf{B}(x_i,r_i))$. Pick η_i , $i = 1,\ldots,n$, such that $\eta_i \in C_0^\infty(\mathbf{B}(x_i,r_i))$ and $\eta_i = 1$ on $\mathbf{B}(x_i,r_i/2)$, and $\eta_0 \in C_0^\infty(\Omega')$ such that $\eta_0 = 1$ on $\Omega' - \bigcup_{i=1}^n \mathbf{B}(x_i,r_i/2)$, in which case $\eta_0(u - f - 2\varepsilon)^+ \in W_0^{1,p}(\Omega)$ by 2.2.9. Also, $(u - f - 2\varepsilon)^+ \leq (u - f(x_i) - \varepsilon)^+ + (-f + f(x_i) - \varepsilon)^+$, so if $\phi = (u - f - 2\varepsilon)^+$, then

$$0 \leq \phi \leq \eta_i(u - f(x_i) - \varepsilon)^+ + \eta_i(-f + f(x_i) - \varepsilon)^+$$

in $\mathbf{B}(x_i,r_i/2)$, $i = 1,\ldots,n$, and

$$0 \leq \phi \leq \eta_0(u - f - 2\varepsilon)^+$$

in $\Omega' - \bigcup_{i=1}^n \mathbf{B}(x_i,r_i/2)$. Let $\psi = \eta_0(u - f - 2\varepsilon)^+ + \sum_{i=1}^n [\eta_i(u - f(x_i) - \varepsilon)^+ + \eta_i(-f + f(x_i) - \varepsilon)^+]$, $0 \leq \phi \leq \psi$ on Ω' and $\psi \in W_0^{1,p}(\Omega')$. $\phi \in W_{loc}^{1,p}(\Omega)$, so $\phi|_{\Omega'} \in W^{1,p}(\Omega')$ since $\bar{\Omega}'$ is compact. Pick $\phi_n \in C^\infty(\Omega') \cap W^{1,p}(\Omega')$ and $\psi_n \in C_0^\infty(\Omega')$ such that $\phi_n \to \phi|_{\Omega'}$ in $W^{1,p}(\Omega')$ and $\psi_n \to \psi$ in $W_0^{1,p}(\Omega')$. Letting $f(x) = x^+$ and using 2.2.4 and 2.2.5, it follows that there exist $f_m \in C^\infty(\mathbf{R})$, $m = 1,2,\ldots$, and a subsequence n_m such that $\{f_m(\phi_{n_m}) - f_m(\phi_{n_m} - \psi_{n_m})\}$ converges in

$W^{1,p}(\Omega')$ to $\phi^+|_{\Omega'} - (\phi|_{\Omega'} - \psi)^+ = \phi|_{\Omega'}$ with gradient $(\chi_{\{\phi>0\}}\nabla\phi - \chi_{\{\phi>\psi\}}(\nabla\phi - \nabla\psi))|_{\Omega'} = (\chi_{\{\phi>0\}}\nabla\phi)|_{\Omega'}$ since $0 < \phi|_{\Omega'} < \psi$. $f_m(\phi_{n_m}) - f_m(\phi_{n_m} - \psi_{n_m}) = 0$ when $\psi_{n_m} = 0$, so is in $C_0^\infty(\Omega')$, therefore $((u-f-2)^+, \chi_{\{u-f-2\varepsilon>0\}}(\nabla u - \nabla f))|_{\Omega'} \in W_0^{1,p}(\Omega')$. By the dominated convergence theorem this converges in $W_0^{1,p}(\Omega')$ to $((u - f)^+, \chi_{\{u>f\}}(\nabla u - \nabla f))|_{\Omega'}$. Doing the above for $-u$, $-f$ instead of u, f shows that $(-(f - u)^+, \chi_{\{u<f\}}(\nabla u - \nabla f))|_{\Omega'} \in W_0^{1,p}(\Omega')$ so that $(u - f, \chi_{\{u\neq f\}}(\nabla u - \nabla f))|_{\Omega'} \in W_0^{1,p}(\Omega')$. ∎

2.3.0 Higher Integrability from Reverse Hölder Inequalities

It will be shown that functions satisfying a maximal function inequality, where lower powers of the function dominate higher powers, actually lie in Higher L^p classes than initially assumed. This is applied in Chapter 3 to prove higher integrability for the gradient of solutions of degenerate elliptic systems.

Theorem 2.3.13 is a generalization of [S1] to weighted spaces, which in turn is an adaptation of a result of F. W. Gehring [GH] to a setting more natural for the analysis of differential equations. The first adaption of [GH] seems to be by M. Giaquinta and G. Modica [GM] although the slightly more general [S1] was done independently, see [GI] for further references.

Only measures which are doubling will be considered here and only very restricted geometries since these are sufficient for applications in Chapter 3. In [S2] Theorem 2.3.13 was proved for very general measures with the restriction that the density of the measure must be uniformly bounded away from zero near the boundary of the domain being considered. This restriction may be weakened even further to the assumption that the measure is doubling near the boundary.

Let Ω be an open set in $\mathbf{R}^d$ and ω a positive Borel measure. It will be assumed that ω is doubling, that is, there exists a constant $c_1 > 0$ such that

$$0 < \omega(B(x,r)) < c_1 \omega B(x,r/2)$$

for all balls $B(x,r) \subseteq \Omega$. By iterating this inequality it is easily seen that there exists a constant $\beta > 0$ such that

$$\omega(B(x,R)) < c_1 \left(\frac{R}{r}\right)^\beta \omega(B(x,r)) \qquad (2.3.1)$$

for all x,r,R, $0 < r < R$, $B(x,r) \subseteq \Omega$.

Let $Q_\infty \subseteq Q_1 \subseteq Q_0 \subseteq \Omega$ be open concentric cubes with sides parallel to the coordinate axes and with side lengths s_∞, $2s_\infty$, $3s_\infty$ respectively. Also for ease in applying the doubling condition it will be assumed that Q_0 lies at least a distance of $15\sqrt{d}\, S_\infty$ from $\partial\Omega$.

L^p norms on Q_∞ will be estimated in terms of L^q norms on Q_0 for some $p > q$. To accomplish this a continuous "iteration" will be carried out on a parameterized collection of cubes Q_t, $1 < t < \infty$, where Q_t is a cube concentric to Q_∞ with sides parallel to Q_∞ and side length $S(t) = S_\infty(1 + t^{-q/\beta})$.

The choice of parameterization is related to the following estimate which can be used to show that a ball centered in some Q_s actually lies totally in Q_t for some specific $t < s$, if its measure is small enough.

Given $B(x,r) \subseteq Q_0$ there exists an $\bar{r} > 0$ such that $r < \bar{r} < 3S_\infty\sqrt{d}$ and $Q_0 \subseteq B(s,\bar{r})$. Using (2.3.1) it now follows that

$$\omega(Q_0) \leq \omega(B(x,\bar{r})) \leq c_1\left(\frac{\bar{r}}{r}\right)^\beta \omega(B(x,r))$$

so that

$$r^\beta \leq c_1(3S_\infty\sqrt{d})^\beta \omega(B(x,r))/\omega(Q_0) \qquad (2.3.2)$$

for all $B(x,r) \subseteq Q_0$.

The maximal functions to be dealt with are defined as follows for $0 < R \leq \infty$.

$$M_R f(x) = \sup\{\omega^{-1}(B) \int |f|\,d\omega : B = B(x,r) \subset \Omega,\ 0 < r < R\}$$

For convenience let $Mf = M_\infty f$. The super-level sets of functions g, f will be of central importance in the main estimate for theorem 2.3.3. These are denoted by $E^*(t) = \{x \in Q_0 : g > t\}$, $E(t) = E^*(t) \cap Q_t$ and $E_*(t) = E^*(t) \cap Q_\infty$. $F^*(t)$, $F(t)$, $F_*(t)$ are defined analogously with respect to f.

<u>2.3.3 Theorem</u>. Suppose g, f are nonnegative Borel measurable functions defined on Q_0, $0 < \alpha < 1$, $b > 1$ and

$$M_R(g^q) \leq bM^q(g) + M(f^q) + \alpha M(g^q) \quad \text{a.e. in } Q_1 \qquad (2.3.4)$$

then there exists a constant p_0 such that if $p_0 > p > q$ then

$$(\frac{1}{\omega(Q_\infty)} \int_{Q_\infty} g^p d\omega)^{1/p} \tag{2.3.5}$$

$$< c[(\frac{1}{\omega(Q_0)} \int_{Q_0} g^q d\omega)^{1/q} + (\frac{1}{\omega(Q_0)} \int_{Q_0} f^p d\omega)^{1/p}]$$

where c depends only on $d,p,q,\alpha,\beta,c_1,b,R/S_\infty$, and p_0 depends only on $d,q,\alpha,c_1,b,R/S_\infty$.

Proof. (2.3.5) will first be proved under the assumption that

$$M_R(g^q) < bM^q(g + f) + \alpha M(g^q) \quad \text{a.e. in } Q_1 \tag{2.3.6}$$

Then either $f \notin L^p(\omega,Q_0)$ in which case 2.3.4 is true or $f \in L^p(\omega,Q_0)$, in which case from propositions 1.1.3, 1.1.4, 1.1.5 and 1.1.9 it follows that $M(M^{1/q}(f^q)) > M^{1/q}(f^q)$ a.e. (2.3.4) then implies that (2.3.6) holds for f replaced by $M^{1/q}(f^q)$ and (2.3.5) follows form propositions 1.1.3 and 1.1.4 and (2.3.5) with f altered as above.

Let

$$\delta = \mathrm{Min}\{(\frac{R}{S_\infty})^\beta \frac{(1 + \alpha)}{2\ 3^\beta d^{\beta/2} c_1 c_5}, \frac{(1 - k^{-q/\beta})^\beta}{2\ 30^\beta d^{\beta/2} c_1}, \frac{\omega(Q_\infty)}{\omega(Q_0)}, 1\} \tag{2.3.7}$$

where c_5 is the constant appearing in 2.3.11 (depends only on c_1,d) and $k^q = 3^q(c_5 b(1 + \alpha)/(1 - \alpha)) > 1$. The doubling condition implies that $\omega(Q_\infty)/\omega(Q_0)$ is bounded below by a positive number depending only on c_1,d so that δ is bounded below by a positive constant depending only on $d,c_1,\alpha,\beta,R/S_\infty$.

Normalize g and f by dividing by

$$\delta((\omega^{-1}(Q_0) \int_{Q_0} g^q d\omega)^{1/q} + (\omega^{-1}(Q_0) \int_{Q_0} f^p d\omega)^{1/p})^{-1}$$

so that without loss of generality (replace g, f by these normalized versions) we may assume that

$$(\omega^{-1}(Q_0) \int_{Q_0} g^q d\omega)^{1/q} + (\omega^{-1}(Q_0) \int_{Q_0} f^p d\omega)^{1/p} = \delta\ . \tag{2.3.8}$$

The remainder of the proof will consist of four parts.

Part I (Decomposition)

Fix $s > k$ and let $t = s/k$. From 2.3.7 it follows that

$$\frac{1}{\omega(Q_s)} \int_{Q_s} g^q d\omega < \frac{\omega(Q_0)}{\omega(Q_s)} \cdot \frac{1}{\omega(Q_0)} \int_{Q_0} g^q d\omega$$

$$< \frac{\omega(Q_0)}{\omega(Q_\infty)} \delta^q < 1 < s^q .$$

Divide Q_s dyadically a minimum number of times so that the subcubes have diameters less than $\mathrm{Min}\{R, S_\infty\}$. For each such Q

$$\frac{1}{\omega(Q)} \int_Q g^q d\omega < \frac{\omega(Q_s)}{\omega(Q)} \cdot \frac{1}{\omega(Q_s)} \int_{Q_s} g^q d\omega < c_3 s^q$$

where c_3 depends only on $\mathrm{Min}\{R,S_\infty\}/S_\infty$, d, c_1. Now subdivide each subcube as in the decomposition lemma of Calderon and Zygmund [ST] to get disjoint subcubes $\{P_i\}$ of Q_s such that

$$g < s \text{ a.e. in } Q_s \setminus (\bigcup_i P_i) \text{ and} \tag{2.3.9}$$

$$s^q < \frac{1}{\omega(P_i)} \int_{P_i} g^q d\omega < c_4 s^q$$

for some c_4 depending only on d, c_1, c_3. The Calderon-Zygmund lemma easily generalizes to the case of a doubling measure because of propositions 1.1.3, 1.1.4, 1.1.5 and 1.1.9. The initial subdivision guarantees that $\mathrm{diam}\, P_i < \mathrm{Min}\{R,S_\infty\}$. Let $G = \bigcup_i P_i$ so

$$\int_G g^q d\omega < c_4 s^q \omega(G) . \tag{2.3.10}$$

Given $x \in P_i$ consider $B(x,r)$ with $r = \mathrm{diam}\, P_i < \mathrm{Min}\{R,S_\infty\}$ so that

$$s^q < \frac{1}{\omega(P_i)} \int_{P_i} g^q d\omega < \frac{\omega(B)}{\omega(P_i)} \left(\frac{1}{\omega(B)} \int_B g^q d\omega\right) < c_5 M_R(g^q)(x) \tag{2.3.11}$$

where c_5 depends only on c_1, d.

Part II (Removal of the term $\alpha M(g^q)$)

Let $F = \{x \in G\colon$ (2.36) holds$\}$ so $\omega(F) = \omega(G)$. If $\alpha = 0$ continue from 2.3.12 otherwise 2.3.11 implies that $c_5^{-1} < s^q c_5^{-1} < M_R(g^q)(x) < M(g^q)(x)$ for all $x \in G$. Given $x \in F$ there exists a ball $B = B(x,r) \subseteq Q_0$ such that

$(1+\alpha)2^{-1}c_5^{-1} < (1+\alpha)2^{-1}M(g^q)(x) < \omega^{-1}(B)\int_B g^q d\omega$ since $(1+\alpha))2^{-1} < 1$. From this it follows that

$$\omega(B) < 2c_5(1+\alpha)^{-1}\int_{Q_0} g^q d\omega < 2c_5(1+\alpha)^{-1}\omega(Q_0)\delta^q .$$

(2.3.2) and (2.3.7) now imply that $r < R$ and so

$$M(g^q)(x) < 2(1+\alpha)^{-1}\omega^{-1}(B)\int_B g^q d\omega < 2(1+\alpha)^{-1}M_R(g^q)(x) .$$

Combining this with (2.3.6) gives

$$M_R(g^q)(x) < \left(\frac{1+\alpha}{1-\alpha}\right) bM^q(g+f)(x) \qquad (2.3.12)$$

for all $x \in F$.

Part III (Basic Estimate)

Given $x \in F$ use (2.3.11), $s = kt$ and the above to get

$$3^q(c_5 b(1+\alpha)/(1-\alpha))t^q = s^q < (c_5 b(1+\alpha)/(1-\alpha))M^q(g+f)(x)$$

so there exists a ball $B = B(x,r) \subseteq Q_0$ such that

$$3t < \frac{1}{\omega(B)}\int (g+f)d\omega . \qquad (2.3.13)$$

It will now be shown that $B \subseteq Q_t$. To see this it is sufficient to show that $r < (S(t) - S(s))/2 = S_\infty t^{-q/\beta}(1 - k^{-q/\beta})/2$. In addition $r < S_\infty/10$ will be proved for later use. The definition of $E^*(t)$ implies that $\int_B g d\omega < t\omega(B) + \int_{B\cap E^*(t)} g d\omega$. Using this, a similar inequality for f and (2.3.13) it follows that

$$t\omega(B) < \int_{B\cap E^*(t)} g d\omega + \int_{B\cap F^*(t)} f d\omega$$

$$< t^{1-q}\int_{E^*(t)} g^q d\omega + t^{1-p}\int_{F^*(t)} f^p d\omega$$

and since $t > 1$ and $p > q$ it is seen that

$$\omega(B) < t^{-q}(\int g^q d\omega + \int f^p d\omega) < t^{-q}(\delta^q + \delta^p)\omega(Q_0) < 2t^{-q}\delta^q\omega(Q_0) .$$

This combined with (2.3.2) and (2.3.7) implies that $r < S_\infty t^{-q/\beta}(1 - k^{-q/\beta})/10$ and so $B \subseteq Q_t$. Since $t > 1$, $k > 1$ it also follows that $r < S_\infty/10$ so that $B(x,5r) \subseteq Q_0$. Using the fact that $B \subseteq Q_t$ and arguing as above it follows that

$$t\omega(B) < \int_{B\cap E(t)} g d\omega + \int_{B\cap F(t)} f d\omega \qquad (2.3.14)$$

F is covered by such balls B. Using the covering lemma 1.1.9 again it is seen that there exists a pairwise disjoint subcollection of balls $\{B_i\}$, $B_i = B(x_i,r_i)$ such that $\{B(x_i,5r_i)\}$ is a cover for F. Now, since $B(x_i,5r_i) \subseteq Q_0$, the doubling condition implies that

$$\omega(G) = \omega(F) \leq \sum \omega(B(x_i,5r_i)) \leq c_i^3 \sum \omega(B(x_i,r_i))$$

and consequently

$$\int_{E(s)} g^q d\omega \leq \int_G g^q d\omega \leq c_4 s^q \omega(G)$$

$$\leq kc_1^3 c_4 s^{q-1}\Big(\int_{E(s/k)} g d\omega + \int_{F(x/k)} f d\omega\Big)$$

from (2.3.10) and (2.3.14).

Part IV (Reduction to Stieltjes integral form)

Let $h(s) = \int_{E(s)} g d\omega$, $H(s) = \int_{F(s)} f d\omega$, so h is nonincreasing and right continuous, and $\operatorname{Lim}_{t\to 0} h(t) = 0$ since $t^{q-1}h(t) \leq \int_{E(t)} g^q d\omega \to 0$ as $t \to 0$. Integrate by parts to get

$$-\int_{(s,\infty)} t^{q-1} dh(t) \tag{2.3.15}$$

$$= (q-1)\int_{(s,\infty)} t^{q-2} h(t) dt + s^{q-1}h(s)$$

$$= (q-1)\int_{(s,\infty)} t^{q-2} \int_{E(t)} g d\omega\, dt + s^{q-1}h(s)$$

$$\leq (q-1)\int_{E(s)} g \int_{(s,g)} t^{q-2} dt\, d\omega + s^{q-1}h(s) \tag{2.3.16}$$

$$= \int_{E(s)} g^q d\omega \tag{2.3.17}$$

$$\leq kc_1^3 c_4 s^{q-1}\Big(\int_{E(s/k)} g d\omega + \int_{F(s/k)} f d\omega\Big).$$

So

$$-\int_{(s,\infty)} t^{q-1} dh(t) \leq kc_1^3 c_4 s^{q-1}(h(s/k) + H(s/k)) \quad \text{for} \quad s > k.$$

Apply lemma 2.3.18 to get

$$-\int_{(k,\infty)} t^{p-1} dh(t)$$

$$\leq c\Big(-\int_{(k,\infty)} t^{q-1} dh(t) + \int_{(k,\infty)} t^{p-2} H(t/k) dt + h(1)\Big).$$

Inequality (2.3.16) is reversed if $E_*(s)$ replaces $E(s)$, so with $s = k$ use (2.3.15) to (2.3.17) (with p replacing q) and the inequality above to get

$$\int_{E_*(k)} g^p d\omega \leq c\Big(\int_{E(k)} g^q d\omega + \int_{(k,\infty)} t^{p-2} \int_{F(t/k)} f d\omega\, dt + h(1)\Big) .$$

$g^p \leq k^{p-q} g^q$ for $g \leq k$ so

$$\int_{Q_\infty} g^p d\omega \leq c(1 + k^{p-q}) \int_{Q_0} g^q d\omega + c \int_{Q_0} f \int_k^{kf} t^{p-2} d\omega + c \int_{E(1)} g d\omega$$

$$\leq c\delta^q \omega(Q_0) + ck^{p-1} \int_{Q_0} f^p d\omega + c \int_{E(1)} g^q d\omega \leq c\omega(Q_\infty)$$

c independent of $\omega(Q_\infty)$. Reversing the normalization of f, g gives (2.3.5).

2.3.18 Lemma. Suppose $h : [1,\infty) \to [0,\infty)$ is nonincreasing, right continuous, and $\lim_{t\to\infty} h(t) = 0$. Also suppose $H : [1,\infty) \to [0,\infty)$ is measurable, $q > 1$, $a > 1$, $k > 1$ and p satisfies $1 > ak^{p-1}(p - q)/(p - 1)$ with $p > q$. If

$$-\int_{(t,\infty)} s^{q-1} dh(s) \leq at^{q-1}(h(t/k) + H(t/k)) \quad \text{for} \quad t \geq k$$

then

$$\int_{(k,\infty)} s^{p-1} dh(s)$$

$$\leq c_1\Big(-\int_{(k,\infty)} s^{q-1} dh(s)\Big) + c_2 \int_{(k,\infty)} t^{p-2} H(t/k) dt + c_3 h(1)$$

Proof. Let $I_p^j = -\int_{(k,j]} s^{p-1} dh(s)$. An integration by parts gives

$$I_p^j = -\int_{(k,j]} t^{p-q} t^{q-1} dh(t) = k^{p-q} I_q^j + (p - q)J , \tag{2.3.19}$$

where $J = \int_{(k,j)} t^{p-q-1}\Big(-\int_{(t,j]} s^{q-1} dh(s)\Big) dt$.

Combining this with the hypothesis it follows that

$$J \leq \int_{(k,j)} t^{p-q-1}\Big(at^{q-1}[h(\tfrac{t}{k}) + H(\tfrac{t}{k})] + \int_{(j,\infty)} s^{q-1} dh(s)\Big) dt \tag{2.3.20}$$

but

$$\int_{(k,j)} t^{p-2}h(\tfrac{t}{k})dt = k^{p-1} \int_{(1,j/k)} t^{p-2}h(t)dt$$

$$= \frac{k^{p-1}}{p-1}\left(\tfrac{j}{k}\, h(\tfrac{j}{k}) - h(1) - \int_{(1,j/k]} t^{p-1}dh(t)\right)$$

$$\leq \frac{k^{p-1}}{p-1}\left(-\int_{(1,j]} t^{p-1}dh(t) - \frac{j^{p-q}}{k^{p-1}} \int_{(j,\infty)} t^{q-1}dh(t)\right)$$

since

$$\left(\tfrac{j}{k}\right)^{p-1} h(\tfrac{j}{k}) = \left(\tfrac{j}{k}\right)^{p-1}\left(h(\tfrac{j}{k}) - h(j)\right) + \left(\tfrac{j}{k}\right)^{p-1} h(j)$$

$$\leq -\int_{(j/k,j]} t^{p-1}dh(t) - \frac{j^{p-q}}{k^{p-1}} \int_{(j,\infty)} t^{q-1}dh(t) .$$

This combined with (2.3.19) and (2.3.20) gives

$$I_p^j \leq k^{p-q} I_q^j + \frac{p-q}{p-1}\, ak^{p-1}\left(-\int_{(1,j]} t^{p-1}dh(t)\right)$$

$$+ \left(\left(\frac{p-q}{p-1}\right)a - 1\right) j^{p-q}\left(-\int_{(j,\infty)} t^{q-1}dh(t)\right)$$

$$- k^{p-q} \int_{(j,\infty)} t^{q-1}dh(t) + (p-q)a \int_{(k,j)} t^{p-2}H(\tfrac{t}{k})dt .$$

Now use that $[a(p-q)/(p-1) - 1] < 0$. Subtract part of the second term from both sides and let $j \to \infty$ to get

$$\left(1 - \left(\frac{p-q}{p-1}\right)ak^{p-1}\right)I_p^\infty \leq k^{p-q} I_q^\infty + (p-q)a \int_{(k,\infty)} t^{p-2}H(\tfrac{t}{k})dt$$

$$+ \frac{p-q}{p-1}\, ak^{p-1}\left(-\int_{(1,k]} t^{p-1}dh(t)\right) ,$$

but $(1 - (p-q)/(p-1))ak^{p-1}) > 0$ by hypothesis, and $-\int_{(1,k]} t^{p-1}dh(t) \leq k^{p-1}h(1)$, so the desired conclusion is reached. ∎

CHAPTER 3

The theme of Chapter 3 is that of establishing continuity for solutions of degenerate elliptic equations.

In Section 3.1.0 both interior and boundary continuity are considered for single equations of the form $\operatorname{div} A(x,u,\nabla u) = B(x,u,\nabla u)$, where A, B satisfy certain natural growth conditions. As a byproduct of this a Harnack inequality is proven for positive solutions.

In Section 3.2.0 estimates are derived for the modulus of continuity of functions in weighted Sobolev spaces, analogous to Morrey's result that functions in $W^{1,p}(\mathbf{R}^d)$, $p > d$, are Hölder continuous. This is relevant since solutions of equations with natural exponent p ($p = 2$ for linear equations) are often contained in such spaces.

In Section 3.3.0, degenerate elliptic systems are considered of the form $\operatorname{div} A_i(x,u,\nabla u) = B_i(x,u,\nabla u)$, $i = 1,\ldots,N$, where A_i, B_i satisfy certain growth conditions. Additional integrability is proven for $|\nabla u|$ and this, combined with the results of Section 3.2.0, establishes continuity in certain borderline cases where 3.2.0 does not apply directly.

In each section an example is worked using equations with degeneracies of the form $\operatorname{dist}^\alpha(x,K)$, for a class of sets K which includes finite unions of C^2 manifolds of co-dimension greater than or equal to 2 (including co-dimension d, i.e. points).

3.1.0 A Harnack Inequality and Continuity of Weak Solutions for Degenerate Elliptic Equations

The main results of this section are a Harnack inequality for positive solutions and the interior and boundary continuity for weak solutions.

The basic structure of the proof of the Harnack inequality is due to Moser [ME1]. Techniques of Trudinger [T1], [T2] are used to replace the John-Nirenberg lemma [JN], which is not of use when the weights are badly degenerate. The proof of the boundary continuity essentially follows that of Gariepy and Ziemer [GZ].

Various results have been proven for linear degenerate equations by Kruzkov [K], Murthy and Stampacchia [MS], P. D. Smith [SM] and Trudinger [T1], [T2] and a degenerate Harnack inequality has been proven by Edmunds and Peletier [EP] for quasi-linear degenerate equations. The present results allow a more general class of degeneracies. The reader should note the related work done independently by E. B. Fabes, D. S. Jerison, C. E. Kenig, and R. P. Serapioni [FKS], [FJK] (see comments preceding 2.2.40).

The equations considered are of the form

$$\operatorname{div} A(x,u,\nabla u) = B(x,u,\nabla u) \,, \tag{3.2.1}$$

where

$$A : \Omega \times \mathbf{R}^1 \times \mathbf{R}^d \to \mathbf{R}^d$$

and

$$B : \Omega \times \mathbf{R}^1 \times \mathbf{R}^d \to \mathbf{R}^1$$

are Borel measurable functions satisfying the conditions

$$|A(x,u,w)| \leq \mu(x)|w|^{p-1} + a_1(x)|u|^{p-1} + a_2(x) , \tag{3.1.2}$$

$$|B(x,u,w)| \leq b_0\lambda(x)|w|^p + b_1(x)|w|^{p-1} + b_2(x)|u|^{p-1} + b_3(x) ,$$

$$A(x,u,w)\cdot w \geq \lambda(x)|w|^p - c_1(x)|u|^p - c_2(x) .$$

$\Omega \subseteq \mathbf{R}^d$ is open, $p > 1$, λ, μ, a_i , $i = 1,2$, b_i , $i = 1,2,3$, c_i , $i = 1,2$, are nonnegative Borel measurable functions on Ω and $\omega = \mu^p\lambda^{-(p-1)}$ and λ are assumed to be integrable with $0 < \lambda \leq \mu < \infty$ almost everywhere.

Identifying ω and λ with the measures they induce, $W^{1,p}(\omega,\lambda,\Omega)$, $W_0^{1,p}(\omega,\lambda,\Omega)$ and $W_{loc}^{1,p}(\omega,\lambda,\Omega)$ will be the Sobolev spaces defined in 2.2.1. For convenience these will be represented by $W^{1,p}_{(\Omega)}$, $W_0^{1,p}(\Omega)$, and $W_{loc}^{1,p}(\Omega)$.

There are a number of useful definitions of weak solution in the present setting. For simplicity a pair $(u,\nabla u) \in W_{loc}^{1,p}(\Omega)$ is called a weak solution of (3.1.1) in an open set V if

$$\int \nabla\Phi\cdot A(x,u,\nabla u) + \Phi B(x,u,\nabla u) = 0 \tag{3.1.3}$$

for all $(\Phi,\nabla\Phi) \in W_0^{1,p}(V)$. In more specific contexts definitions such as that in [T1] may be more natural. In any case the basic methods are quite flexible in adapting to different definitions of weak solution.

The following Sobolev inequalities will be assumed. $B = B(x_0,r)$ is a ball of radius r, $\bar{B} \subseteq \Omega$, and x_0 , r vary depending on the specific result being considered.

$$\left(\frac{1}{\omega(B)} \int_B |\Phi|^q\omega\right)^{1/q} \leq rs(r)\left(\frac{1}{\omega(B)} \int_B |\nabla\Phi|^p\lambda\right)^{1/p} \tag{3.1.4}$$
$$+ t(r)\left(\frac{1}{\omega(B)} \int_B |\Phi|^p\omega\right)^{1/p}$$

for all $(\Phi,\nabla\Phi) \in W_0^{1,p}(B)$, where $q > p$ and $s(r) \geq 2$ (for computational simplicity);

$$\int_B \left|\Phi - \frac{1}{\omega(B)} \int_B \Phi\omega\right|^p\omega \leq r^p p(r) \int_B |\nabla\Phi|^p\lambda + q(r)\omega(B) \tag{3.1.5}$$

for all $(\Phi,\nabla\Phi) \in W^{1,p}(B(x_0,\bar{r}))$, and some $\bar{r} > r$;

$$\int_B |\Phi|^p F_r \le \varepsilon r^p s_0 \int_B |\nabla\Phi|^p \lambda + \varepsilon^{-\delta} s_F(r) \int_B |\Phi|^p \omega \tag{3.1.6}$$

for all $(\Phi,\nabla\Phi) \in W_0^{1,p}(B)$ and $0 < \varepsilon \le 1$, where s_0 is either 0 or 1 and F_r will be defined slightly differently in each of Theorems 3.1.10, 3.1.15, and Corollaries 3.1.12, 3.1.13.

The weights ω, λ for which (3.1.4) and (3.1.5) hold, with $t(r) = q(r) = 0$ and $\Phi \in C_0^\infty(B)$ and $C_0^\infty(B(x_0,\bar{r}))$, respectively, are characterized in Theorem 2.2.41. Simple limit procedures as done in Lemma 3.1.7 then show that (3.1.4) and (3.1.5) hold for general Sobolev functions.

If $F_r \le s_F(r)\omega$, then (3.1.6) is trivially true with $s_0 = 0$; otherwise (3.1.6) can be deduced from inequalities such as (3.1.8), for which the weights have been characterized in Theorem 2.2.41.

<u>3.1.7 Lemma</u>. Assume for some s, $1 \le s < p$, that

$$\int_B |\Phi|^s F_r \le c(r) r^s \int_B |\nabla\Phi|^s \lambda^{s/p} \omega^{(p-s)/p} \tag{3.1.8}$$

for all $\Phi \in C_0^\infty(B)$. Then

$$\int_B |\Phi|^p F_r \le \varepsilon r^p \int |\nabla\Phi|^p \lambda \tag{3.1.9}$$
$$+ c(p,s)\varepsilon^{-s/(p-s)} c^{p/(p-s)}(r) \int |\Phi|^p \omega$$

for all $(\Phi,\nabla\Phi) \in W_0^{1,p}(B)$.

<u>Proof</u>. Given $u \in C_0^\infty(B)$, let $\Phi = u^{p/s}$. $\nabla\Phi = \frac{p}{s} u^{(p-s)/s}\nabla u$, so by (3.1.8)

$$\int_B |u|^p F_r \le \left(\frac{p}{s}\right)^s c(r) r^s \int_B |u|^{p-s} |\nabla u|^s \lambda^{s/p} \omega^{(p-s)/p} .$$

Use Younges' inequality to show that

$$\left(\frac{p}{s}\right)^s c(r) r^s |u|^{p-s} |\nabla u|^s \lambda^{s/p} \omega^{(p-s)/p}$$
$$\le C(p,s)\varepsilon^{-s/(p-s)} c^{p/(p-s)}(r) |u|^p \omega + \varepsilon r^p |\nabla u|^p \lambda$$

so that (3.1.9) is true for $u \in C_0^\infty(B)$. Given $(\psi,\nabla\psi) \in W_0^{1,p}(B)$, pick $\phi_n \in C_0^\infty(B)$ such that $(\phi_n,\nabla\phi_n) \to (\psi,\nabla\psi)$ in $W_0^{1,p}(\Omega)$. Using (3.1.9) with $\Phi = \phi_n - \phi_m$ it is seen that $\{\phi_n\}$ is Cauchy in

$L^p(F_r,B)$ and since $\phi_{n_i} \to \psi$ a.e. for some subsequence $\{n_i\}$ it follows that $\psi \in L^p(F_r,B)$ and $\phi_n \to \psi$ in $L^p(F_r,B)$. Letting $\Phi = \phi_n$ in (3.1.9) and letting $n \to \infty$ shows that (3.1.9) is true for $(\Phi,\nabla\Phi) = (\psi,\nabla\phi)$, as required. ∎

For Theorem 3.1.10 it will be assumed that $a_2 = b_3 = c_2 = 0$, $F_R = R^p[c_1 + b_1^p\lambda^{-(p-1)} + b_2 + a_1^{p/(p-1)}\mu^{-p/(p-1)}\lambda]$, $\overline{B(x_0,R)} \subseteq \Omega$, and (3.1.4), (3.1.5), and (3.1.6) hold for $r = R$.

<u>3.1.10 Theorem</u>. If u is a positive weak solution of (3.1.1) in $B(x_0,R)$ with $u \leq M < \infty$ and $0 < \theta < 1$, then

$$\sup_{B(x_0,\theta R)} u \leq C(R) \inf_{B(x_0,\theta R)} u \tag{3.1.11}$$

for

$$C(R) = c[(s(R)+t(R))\exp(p(R)H(R)+q(r))]^{c(s(R)(s_F^{1/p}(R)+1)+t(R))^{q/(q-p)}}$$

and

$$H(R) = 1 + \omega^{-1}(B(x_0,R)) \int_{B(x_0,R)} [(c_1 + b_1^p\lambda^{-(p-1)} + b_2)R^p + a_1R^{p-1}] ,$$

c depending only on p, q, θ, M, b_0.

The proofs of Theorem 3.1.10 and the following results will be deferred until later.

<u>Remarks</u>.

The boundedness assumption for weak solutions is not essential if $b_0 = 0$. In this case methods of Aronson and Serin [AS] can be used to achieve similar results.

As in [T3] the Harnack inequality may be split into two parts, one relevant to subsolutions and one to supersolutions.

The John-Nirenberg lemma [JN] generalizes easily to accomodate doubling measures, that is, measures μ such that $\mu(B(x,2r)) \leq c\mu(B(x,r))$. So if integration against ω is a doubling measure and if the inequalities (3.1.4), (3.1.5), and (3.1.6) hold for $0 < r \leq R$ with $p(r)$, $q(r)$, $H(r)$ bounded, then the "crossover" can be done as in [T3].

The last remark applies to supersolution calculations as well, but if full solutions alone are of concern and if the remaining Sobolev constants $s(r)$, $t(r)$, $s_F(r)$ are bounded for $0 < r \leq R$, then a simple method due to Bombieri [BI] and appearing in [ME2] can be used as well, to prove (3.1.11) with R replaced by r and $C(r)$ bounded.

If, as in the symmetric linear case, $|w_1 \cdot A(x,u,w_2)| \le |w_1 \cdot A(x,u,w_1)||w_2 \cdot A(x,u,w_2|$, then the derivation of the fundamental inequality in the proof of Theorem 3.1.10 may be improved as in [ME2]. This leads to the replacement of ω by the smaller weight μ, thus allowing consideration of more degenerate weights.

Definition. If $x \in \Omega$ and $\lim_{r \to 0} \inf_{B(x,r)} V = \inf_{\Omega} V$, then it is said that V achieves its essential minimum at x. The analogous definition is adopted for essential maximum at x.

3.1.12 Corollary. Assume $0 = a_1 = a_2 = b_2 = b_3 = c_1 = c_2$, $F_r = r^p b_1^p \lambda^{-(p-1)}$, Ω is open and connected and for all $x \in \Omega$, $\exists\ r > 0$ such that inequalities (3.1.4), (3.1.5), and (3.1.6) hold for $x_0 = x$ and $\overline{B(x,r)} \subseteq \Omega$.

If V is a bounded weak solution of (3.1.1) in Ω and V achieves its essential minimum or maximum at an interior point, then V is constant (off a set of measure zero).

Remarks. If a_1, a_2, b_2, b_3, c_1, c_2 are not assumed to be zero, then a weak maximum principle may be proven similar to that of [AS].

If $C(R)$ in (3.1.11) depends on R in an appropriate manner, then a Liouville theorem may be proven as in [ME1].

In Corollary 3.1.13 it is shown that a slightly altered Harnack inequality holds if a_2, b_3, c_3 are not zero. The function $K(r)$ is usually chosen to be r^α for some $\alpha > 0$.

Let $F_R = R^p[c_1 + c_2(K^{-p}(R) + b_1^p\lambda^{-(p-1)} + b_2 + b_3K^{-(p-1)}(R) + (a_1 + a_2K^{-(p-1)}(R))^{p/(p-1)}\mu^{-p/(p-1)}\lambda]$ and assume (3.1.4), (3.1.5), and (3.1.6) for $r = R$.

3.1.13 Corollary. If u is a positive weak solution of (3.1.1) in $B(x_0,R)$ with $u \le M$ and $0 < \theta < 1$, then

$$\sup_{B(x_0,\theta R)} u \le C(R) \inf_{B(x_0,\theta R)} u + (C(R) - 1)K(R) , \tag{3.1.14}$$

where $C(R)$ is as in Theorem 3.1.10.

For Theorem 3.1.15 let $F_r = r^p[(c_1M + c_2)K^{-p}(r) + b_1^p\lambda^{-(p-1)} + (b_2M + b_3)K^{-(p-1)}(r) + (a_1M + a_2)^{p/(p-1)}K^{-p}(r)\mu^{-p/(p-1)}\lambda]$, and assume (3.1.4), (3.1.5), and (3.1.6) for $0 < r \le R$.

3.1.15 Theorem. If u is a weak solution of (3.1.1) in $B(x_0,R)$ with $|u| \le M/2$ and $C(R)$, $K(R)$, θ, M as in Corollary 3.1.13, then

$$\lim_{k\to\infty} \operatorname*{Osc}_{B(x_0,r_k)} u = 0 \quad \text{if} \quad \sum_{k=0}^{\infty} C^{-1}(r_k) = \infty \quad \text{and} \quad \lim_{k\to\infty} C(r_k)K(r_k) = 0,$$

where $r_k = \theta^k R$.

If $C(r_k)$ is bounded and $K(r_k) \leq c' r_k^{\alpha'}$ for some c', $\alpha' > 0$, then

$$\operatorname*{Osc}_{B(x_0,r_k)} u \leq c r_k^{\alpha} \quad \text{for some } c,\ \alpha > 0 .$$

These conditions are sharp in the sense of Lemma 3.1.16.

In addition, if $C(r)$ is nondecreasing and $K(r)$ is nonincreasing as $r \to 0$, then

$$\operatorname*{Osc}_{B(x_0,r_k)} u \leq e^{-cg(r_k)} \Big(\operatorname*{Osc}_{B(x_0,R)} u + c \int_0^{g(r_k)} \gamma \circ g^{-1}(t) e^{ct} dt \Big) ,$$

where $g(s) = \int_s^R \frac{1}{C(r)} \frac{dr}{r}$ and $\gamma(s) = C(s)K(s/\theta)$.

Remark. Semicontinuity results for subsolutions and supersolutions may be proven as in [T1] using the calculations mentioned in the second remark after Theorem 3.1.10.

Example.

Let K be as in Theorem 2.2.56 such that 2.2.61-3 hold and $\omega(x) = \mu(x) = \lambda(x) = \operatorname{dist}^{\alpha}(x,K)$, $\alpha > -\gamma$. From this and a limiting argument (as in Lemma 3.1.7) it follows for some $q > p$ and all $B(x_0,r) \subseteq \mathbf{R}^d$, that (3.1.4) and (3.1.5) hold with $s(r) = p(r) = 1$ and $t(r) = q(r) = 0$. For simplicity assume that a_1, a_2, c_1, c_2, b_1, b_2, b_3 are bounded by a constant multiple of ω and choose $k(r) = r + r^{(p-1)/p}$. This implies that $F_r \leq c\omega$, so (3.1.6) is trivially true with $s_0 = 0$ and $s_F(r) = 1$. Also $C(r)$ is bounded for $r < R < \infty$. It now follows from Theorem 3.1.15 that if u is a bounded weak solution of (3.1.1) in Ω, then u is locally Hölder continuous.

3.1.16 Lemma. Let $r_k = \theta^k R$ for $0 < \theta < 1$, $R > 0$. Assume

$$C(r_k) \geq \delta > 1 , \tag{3.1.17}$$

and

$$\operatorname*{Osc}_{B(x_0,r_k)} u \leq a_{k-1}\Big(\operatorname*{Osc}_{B(x_0,r_{k-1})} u + 2K(r_{k-1})\Big) \tag{3.1.18}$$

for $k = 0,1,2,\ldots,$ where $a_k = \frac{C(r_k) - 1}{C(r_k) + 1}$. Then

$$\operatorname*{Osc}_{B(x_0,r_n)} u \leq \Big(\prod_{k=0}^{n-1} a_k\Big) \operatorname*{Osc}_{B_R} u + 2 \sum_{j=0}^{n-1} k(r_j) \sum_{k=j}^{n-1} a_k , \tag{3.1.19}$$

and if

$$\sum_{k=0}^{\infty} C^{-1}(r_k) = \infty \quad \text{and} \quad C(r_k)K(r_k) \to 0 \tag{3.1.20}$$

as $k \to \infty$, then

$$\operatorname*{Osc}_{B(x_0,r_n)} u \to 0 \quad \text{as} \quad n \to \infty . \tag{3.1.21}$$

(3.1.22) This is sharp because if (3.1.20) does not hold, then there exists u such that $\operatorname*{Osc}_{B(x_0,r_n)} u \geq \lambda > 0$.

If $C(r_k)$ is bounded and $K(r_k) \leq cr_k^{\alpha}$ for some c, $\alpha > 0$, then

$$\operatorname*{Osc}_{B(x_0,r_n)} u \leq c' r_n^{\alpha'} \quad \text{for some } c', \ \alpha' > 0 . \tag{3.1.23}$$

If $C(r_k)$ is nondecreasing as $k \to \infty$, then this is sharp as well.

If $C(r)$ is nondecreasing and $K(r)$ is nonincreasing as $r \to 0$, then

$$\operatorname*{Osc}_{B(x_0,r_n)} u \leq e^{-g(r_n)}\Big(\operatorname*{Osc}_{B(x_0,R)} u + 2c \int_0^{g(r_n)} \gamma \circ g^{-1}(t) e^{ct} dt\Big) ,$$

where $g(s) = \int_S^R \frac{1}{C(r)} \frac{dr}{r}$, $\gamma(s) = C(s)K(s/\theta)$ and $c = (\log \theta^{-1})^{-1}$.

Proof of Theorem 3.1.10. The fundamental inequality (3.1.29) is proven; then this is iterated to give (3.1.31). The final step is the crossover from L^p norms of u with $p > 0$ to those with $p < 0$. This is accomplished by iterating norms of $\log u$.

It can be assumed without loss of generality that u is strictly positive, otherwise let $\bar{u} = u + \varepsilon$, $\varepsilon > 0$. $\bar{u}$ is a weak solution of $\operatorname{div} \bar{A} = \bar{B}$, where $\bar{A}(x,\bar{u},\nabla\bar{u}) = A(x,\bar{u} - \varepsilon,\nabla\bar{u})$ and $B(x,\bar{u},\nabla\bar{u}) = B(x,\bar{u} - \varepsilon,\nabla\bar{u})$, and since $\bar{A}$, $\bar{B}$ satisfy (3.1.2), the following proof gives $\sup_{B(x_0,\theta R)} \bar{u} \leq C(R) \inf_{B(x_0,\theta R)} \bar{u}$. (3.1.11) is recovering by letting $\varepsilon \to 0$.

Throughout the proof, c will represent a constant depending only on p, q, θ, M, b_0, and will change from time to time.

Let $\Phi = \phi^p u^\beta \exp(b_0 \operatorname{sign}\beta\, u)$, $\beta \neq 0$, $\phi \in C_0^\infty(B(x_0,R))$. Several applications of Proposition 2.2.2 show that $\Phi \in W_0^{1,p}(B(x_0,R))$ with

$$\nabla\Phi = p\phi^{p-1}\nabla\phi u^\beta \exp(b_0 \operatorname{sign}\beta\, u) + \beta\phi^p u^{\beta-1} u \exp(b_0 \operatorname{sign}\beta\, u)$$

$$+ b_0 \operatorname{sign}\beta\, \phi^p u^\beta \exp(b_0 \operatorname{sign}\beta\, u)\nabla u ,$$

and so

$$\beta \int \phi^p u^{\beta-1}\exp(b_0 \operatorname{sign}\beta\, u)\nabla u\cdot A = -\int \phi^p u^\beta \exp(b_0 \operatorname{sign}\beta\, u)B$$

$$- p \int \phi^{p-1} u^\beta \exp(b_0 \operatorname{sign}\beta\, u)\nabla\phi\cdot A$$

$$- b_0 \operatorname{sign}\beta \int \phi^p u^\beta \exp(b_0 \operatorname{sign}\beta\, u)\nabla u\cdot A .$$

Now multiply by $\operatorname{sign}\beta$, use the structure inequalities (3.1.2) and $u \leq M$, and let $E = \exp(b_0 \operatorname{sign}\beta\, u)$ to get

$$|\beta| \int \phi^p u^{\beta-1} |\nabla u|^p \lambda E \tag{3.1.24}$$

$$\leq |\beta| \int \phi^p u^{\beta+p-1} c_1 E + b_0 \int \phi^p u^\beta |\nabla u|^p \lambda E$$

$$+ \int \phi^p u^\beta |\nabla u|^{p-1} b_1 E + \int \phi^p u^{\beta+p-1} b_2 E$$

$$+ p \int \phi^{p-1}|\nabla\phi| u^\beta |\nabla u|^{p-1} \mu E + p \int \phi^{p-1}|\nabla\phi| u^{\beta+p-1} a_1 E$$

$$- b_0 \int \phi^p u^\beta |\nabla u|^p \lambda E + b_0 M \int \phi^p u^{\beta+p-1} c_1 E .$$

The second and the seventh terms on the right-hand side cancel. This is in fact the reason for introducing $\exp(b_0 \operatorname{sign}\beta\, u)$ in the test function Φ.

The following inequalities are proven using Young's inequality.

$$\begin{cases} u^\beta |\nabla u|^{p-1} b_1 \leq \varepsilon_1^{-(p-1)} u^{\beta+p-1} b_1^p \lambda^{-(p-1)} \\ \qquad\qquad + \dfrac{(p-1)}{p}\, \varepsilon_1 u^{\beta-1} |\nabla u|^p \lambda , \\ \phi^{p-1}|\nabla\phi| u^\beta |\nabla u|^{p-1}\mu \leq \dfrac{\varepsilon_2^{-(p-1)}}{p} |\nabla\phi|^p u^{\beta+p-1} \mu^p \lambda^{-(p-1)} \\ \qquad\qquad + \dfrac{(p-1)}{p}\, \varepsilon_2 \phi^p u^{\beta-1} |\nabla u|^p \lambda , \end{cases} \tag{3.1.25}$$

$$\phi^{p-1}|\nabla\phi|a_1 \le \frac{1}{p}|\nabla\phi|^p\mu^p\lambda^{-(p-1)} + \frac{(p-1)}{p}\phi^p a_1^{p/(p-1)}\mu^{-p/(p-1)}\lambda \ . \qquad (3.1.26)$$

Applying these to (3.1.24) with $\varepsilon_1 = \frac{p}{p-1}\cdot\frac{|\beta|}{4}$ and $\varepsilon_2 = \frac{|\beta|}{4(p-1)}$, and absorbing the gradient terms into the left-hand side it follows that

$$|\tfrac{\beta}{2}|\int \phi^p u^{\beta-1}|\nabla u|^p\lambda E$$

$$\le |\beta|\int \phi^p u^{\beta+p-1}c_1 E + \frac{\varepsilon_1^{-(p-1)}}{p}\int \phi^p u^{\beta+p-1}b_1^p\lambda^{-(p-1)}E$$

$$+ \int \phi^p u^{\beta+p-1}b_2 E + (\varepsilon_2^{-(p-1)} + 1)\int |\nabla\phi|^p u^{\beta+p-1}\mu^p\lambda^{-(p-1)}E$$

$$+ \int \phi^p u^{\beta+p-1}a_1^{p/(p-1)}\mu^{-p/(p-1)}\lambda E + b_0 M\int \phi^p u^{\beta+p-1}c_1 E \ .$$

Considering that $1 \le e^{b_0 u} \le e^{b_0 M}$ and $u^{\beta-1}|\nabla u|^p = (\frac{p}{\gamma})^p|\nabla u^{\gamma/p}|^p$ for $\gamma = \beta + p - 1 \ne 0$, it follows for $\gamma \ne p-1$, $\gamma \ne 0$, that

$$R^p\int \phi^p|\nabla u^{\gamma/p}|^p\lambda \le cc(\gamma)\int [R^p|\nabla\phi|^p\omega + \phi^p F_R]u^\gamma \ ,$$

where $F_R = R^p[c_1 + b_1^p\lambda^{-(p-1)} + b_2 + a_1^{p/(p-1)}\mu^{-p/(p-1)}\lambda]$ and $c(\gamma) = (1 + |\beta|^{-p})\gamma^p$.

Now use (3.1.6) with $\Phi = \phi u^{\gamma/p}$ and $\varepsilon^{-1} = \max\{1, 2cc(\gamma)\}$. The resulting gradient terms may be absorbed on the left to give

$$R^p\int \phi^p|\nabla u^{\gamma/p}|^p\lambda \le cc(\gamma)\int (\varepsilon^{-\delta}s_F(R)\phi^p + R^p|\nabla\phi|^p)u^\gamma\omega \qquad (3.1.27)$$

(recall $\lambda \le \omega$). Using inequality (3.1.4) with $\Phi = \phi u^{\gamma/p}$ and $B_R = B(x_0,R)$, it follows that

$$\left(\frac{1}{\omega(B_R)}\int_{B_R} \phi^q u^{\gamma\cdot q/p}\omega\right)^{1/q} \le c\left(\frac{1}{\omega(B_R)}\int_{B_R} G u^\gamma\omega\right)^{1/p} ,$$

with $G = s^p(R)C(\gamma)(\varepsilon^{-\delta}s_F(R)\phi^p + R^p|\nabla\phi|^p) + s^p(R)R^p|\nabla\phi|^p + t^p(R)\phi^p$.

<u>3.1.28</u>. This inequality is now iterated. To do this choose θ_i, $i = 0,1,2,3$, such that $\frac{\rho}{R} = \theta_0 < \theta_1 < \theta_2 < \theta_3 = 1$, and let $\rho_i = R\theta_i$. Also let $B_k = B(x_0,r_k)$, where $r_k = R(\theta_0 + (\theta_1 - \theta_0)2^{-k})$. Choose $\phi_k \in C_0^\infty(B_k)$ such that $0 \le \phi_k \le 1$, $\phi_k = 1$ on B_{k+1} and

$|\nabla\phi_k| < 2^{k+2}/[R(\theta_1 - \theta_2)]$, and let $\gamma_k = \gamma_0(q/r)^k$. with these choices of ϕ, γ it follows that

$$\left(\frac{1}{\omega(B_R)}\int_{B_{k+1}} u^{\gamma_{k+1}}\omega\right)^{1/|\gamma_{k+1}|} < C_{1,k}(R)\left(\frac{1}{\omega(B_R)}\int_{B_k} u^{\gamma_k}\omega\right)^{1/|\gamma_k|} \tag{3.1.29}$$

where

$$C_{1,k}(R) = c\Big(s^p(R)\Big[C(\gamma_k)\Big(\varepsilon_k^{-\delta}s_F(R) + \left(\frac{2^{k+2}}{\theta_1 - \theta_0}\right)^p\Big) + \left(\frac{2^{k+2}}{\theta_1 - \theta_0}\right)^p\Big] + t^p(R)\Big)^{1/|\gamma_k|} .$$

$C(\gamma_k)$ and in turn $C_{1,k}(R)$ blow up if γ_k tends to $p - 1$. If $\gamma_0 < 0$, then this is impossible. If $\gamma_0 > 0$, then γ_0 must be chosen carefully.

3.1.30. Given $1 > \sigma > 0$ (to be determined in 3.1.38), pick γ_0 such that $\frac{p}{q}\sigma < \gamma_0 < \sigma$ and $|\beta_k| = |\gamma_k - (p - 1)| > \left(\frac{p-1}{2}\right)\left(1 - \frac{p}{q}\right)$ for all k. With this choice of γ_0, $C(\gamma) < c\gamma_k^p$, and so after a crude calculation it follows that

$$C_{1,k}(R) < c^{k/|\gamma_k|}\Big(s(R)(|\gamma_0|s_F^{1/p}(R) + 1) + t(R)\Big)^{p/|\gamma_k|} .$$

Iterating (3.1.29) it follows that

$$\sup_{B_\rho} u^{\operatorname{sign}\gamma_0} < \left(C_1(R)\frac{1}{\omega(B_R)}\int_{B_{\rho_1}} u^{\gamma_0}\omega\right)^{1/|\gamma_0|}, \tag{3.1.31}$$

with $C_1(R) = \prod_{k=0}^{\infty} C_{1,k}(R)$. Recalling that $\gamma_k = \gamma_0\left(\frac{q}{p}\right)^k$, it follows that

$$C_1(R) < c\Big(s(R)(|\gamma_0|s_F^{1/p}(R) + 1) + t(R)\Big)^{pq/(|\gamma_0|(q-p))} .$$

The last step in the proof is the "crossover". First an inequality is derived which in the uniformly elliptic case leads to the conclusion that $\log u \in$ B.M.O. which, in turn, gives the "crossover". If ω is not a doubling measure, this is not sufficient, and one further iterative procedure is necessary to get the "crossover".

3.1.32. Take $\beta = 1 - p$ at (3.1.24), and proceed as before but without using inequality (3.1.26) to get

$$\int \phi^p |\nabla \log u|^p \lambda \leq c \int \phi^p (c_1 + b_1^p \lambda^{-(p-1)} + b_2) + \phi^{p-1}|\nabla\phi| a_1 + |\nabla\phi|^p \omega \ .$$

Choose $\phi \in C_0^\infty(B(x_0,R))$ such that $0 \leq \phi \leq 1$, $\phi = 1$ on B_{ρ_2} and $|\nabla\phi| \leq 2/(R - \rho_2) = R^{-1}(2/(1 - \theta_2))$, so

$$\rho_2^p \int_{B_{\rho_2}} |\nabla \log u|^p \lambda \leq cH(R)\omega(B_R) \ ,$$

where

$$H(R) = 1 + \frac{1}{\omega(B_R)} \int_{B_R} [(c_1 + b_1^p \lambda^{-(p-1)} + b_2)R^p + a_1 R^{p-1}] \ .$$

Using inequality (3.1.5) it follows that

$$\int_{B_{\rho_2}} |\log \frac{u}{k}|^p \omega \leq cK(R)\omega(B_R) \ , \tag{3.1.33}$$

where $K(R) = p(R)H(R) + q(R)$ and $k = \frac{1}{\omega(B_{\rho_2})} \int_{B_{\rho_2}} (\log u)\omega$.

To derive the inequality needed in the final iteration let the test function be

$$\Phi = \eta^p u^{1-p}(|v|^\beta + (\frac{p\beta}{p-1})^\beta)\exp(-b_0 u) \ ,$$

where $\eta \in C_0^\infty(B_R)$, $\eta \geq 0$, $v = \log \frac{u}{k}$, k as above, and $\beta \geq 1$. Repeated applications of Proposition 2.2.2 show that $\Phi \in W_0^{1,p}(B(x_0,R))$ with

$$\nabla\Phi = p\eta^{p-1}\nabla\eta u^{1-p}(|v|^\beta + (\frac{p\beta}{p-1})^\beta)\exp(-b_0 u)$$

$$- \eta^p u^{-p}\nabla u((p-1)(|v|^\beta + (\frac{p\beta}{p-1})^\beta) - \beta|v|^{\beta-1}\mathrm{sign}\ v)\exp(-b_0 u)$$

$$- b_0\eta^p u^{1-p}(|v|^\beta + (\frac{p\beta}{p-1})^\beta)\exp(-b_0 u)\nabla u \ .$$

<u>3.1.34.</u> Substituting this in 3.1.3 and letting $E = \exp(-b_0 u)$, it follows that

$$\int \eta^p u^{-p} E((p-1)(|v|^\beta + (\frac{p\beta}{p-1})^\beta) - \beta|v|^{\beta-1}\mathrm{sign}\ v)\nabla u\cdot A$$

$$= p\int \eta^{p-1}u^{1-p}E(|v|^\beta + (\frac{p\beta}{p-1})^\beta)\nabla\eta\cdot A$$

$$- b_0 \int \eta^p u^{1-p} E(|v|^\beta + (\frac{p\beta}{p-1})^\beta) \nabla u \cdot A$$

$$+ \int \eta^p u^{1-p} E(|v|^\beta + (\frac{p\beta}{p-1})^\beta) B \ .$$

Use the structure inequalities (3.1.2) and $\frac{p\beta}{p-1} |v|^{\beta-1} \leq |v|^\beta + (\frac{p\beta}{p-1})^\beta$ to get

$$\frac{(p-1)^2}{p} \int \eta^p u^{-p} |\nabla u|^p E(|v|^\beta + (\frac{p\beta}{p-1})^\beta) \lambda$$

$$\leq p \int \eta^p E(|v|^\beta + (\frac{p\beta}{p-1})^\beta) c_1$$

$$+ p \int \eta^{p-1} |\nabla\eta| u^{1-p} E(|v|^\beta + (\frac{\beta p}{p-1})^\beta) |\nabla u|^{p-1} \mu$$

$$+ p \int \eta^{p-1} |\nabla\eta| E(|v|^\beta + (\frac{\beta p}{p-1})^\beta) a_1$$

$$- b_0 \int \eta^p u^{1-p} E(|v|^\beta + (\frac{\beta p}{p-1})^\beta) |\nabla u|^p \lambda$$

$$+ b_0 M \int \eta^p E(|v|^\beta + (\frac{p\beta}{p-1})^\beta) c_1$$

$$+ b_0 \int \eta^p u^{1-p} E(|v|^\beta + (\frac{p\beta}{p-1})^\beta) |\nabla u|^p \lambda$$

$$+ \int \eta^p u^{1-p} E(|v|^\beta + (\frac{p\beta}{p-1})^\beta) |\nabla u|^{p-1} b_1$$

$$+ \int \eta^p E(|v|^\beta + (\frac{p\beta}{p-1})^\beta) b_2 \ .$$

<u>3.1.35</u>. The fourth and sixth terms cancel. Eliminate E, multiply by R^p, use Young's inequality as before, and recall that $\nabla v = u^{-1} \nabla u$ to get

$$R^p \int \eta^p |\nabla v|^p (|v|^\beta + (\frac{p\beta}{p-1})^\beta) \lambda$$

$$\leq c \int (|v|^\beta + (\frac{p\beta}{p-1})^\beta)(\eta^p F_R + R^p |\nabla\eta|^p \omega) \ .$$

Use $|v|^\beta \leq (|v|^{\gamma/p} + (\frac{p\beta}{p-1})^{\beta/p})^p$, $\gamma = \beta + p - 1$ $(\beta > 1)$ on the right and $\beta |v|^{\beta-1} \leq |v|^\beta + (\frac{p\beta}{p-1})^\beta$ on the left to get

$$R^p \int \eta^p |\nabla v^{\gamma/p}|^p \lambda$$

$$\leq c\gamma^p \int (|v|^{\gamma/p} + (\frac{p\beta}{p-1})^{\beta/p})^p (\eta^p F_R + R^p |\nabla\eta|^p \omega) \ .$$

Use inequality (3.1.6) on the F_R term with $\Phi = \eta(|v|^{\gamma/p} + (\frac{p\beta}{p-1})^{\beta/p})$ and $\varepsilon^{-1} = 2c\gamma^p > 1$ and then cancel gradient terms to get

$$R^p \int \eta^p |\nabla v^{\gamma/p}|^p \lambda \le c\gamma^p \int (\varepsilon^{-\delta} s_F(R)\eta^p |\nabla\eta|)(|v|^{\gamma/p} + (\frac{p\beta}{p-1})^{\beta})^p \omega .$$

Use inequality (3.1.4) with $\Phi = \eta v^{\gamma/p}$ to get

$$(\frac{1}{\omega(B_R)} \int \eta^q |v|^{\gamma q/p} \omega)^{1/q}$$

$$\le s(R)(\frac{c\gamma^p}{\omega(B_R)} \int (\varepsilon^{-\delta} s_F(R)\eta^p + R^p|\nabla\eta|^p)(|v|^{\gamma} + (\frac{p\beta}{p-1})^{\beta})\omega$$

$$+ \frac{1}{\omega(B_R)} \int R^p |\nabla\eta|^p |v|^{\gamma}\omega)^{1/p} + t(R)(\frac{1}{\omega(B_R)} \int \eta^p |v|^{\gamma}\omega)^{1/p}$$

$$\le c(\frac{1}{\omega(B_R)} \int (s(R)\gamma^p(\varepsilon^{-\delta} s_F(R)\eta^p + R^p|\nabla\eta|^p) + t^p(R)\eta^p) \times$$

$$(|v|^{\gamma} + (\frac{p\beta}{p-1})^{\beta}))^{1/p} .$$

Let $B_k = B(x,r_k)$, $r_k = R(\theta_1 - (\theta_2 - \theta_1)2^{-k})$, $k = 0,1,\ldots$. Pick $\eta_k \in C_0^\infty(B_k)$ such that $0 \le \eta_k \le 1$, $\eta_k = 1$ on B_{k+1}, and $|\nabla\eta_k| \le 2^{k+2}/(R(\theta_2 - \theta_1))$, and let $\gamma_k = p(q/p)^k$. This gives

$$(\frac{1}{\omega(B_R)} \int_{B_{k+1}} |v|^{\gamma_{k+1}}\omega)^{1/\gamma_{k+1}}$$

$$\le C_{2,k}(R)(\frac{1}{\omega(B_R)} \int_{B_k} |v|^{\gamma_k} + (\frac{p\beta_k}{p-1})^{\beta_k}\omega)^{1/\gamma_k} ,$$

with $C_{2,k}(R) = c(s^p(R)\gamma_k^p(\varepsilon_k^{-\delta} s_F(R) + (\frac{2^{k+2}}{\theta_2 - \theta_1})^p) + t^p(R))^{1/\gamma_k}$.
Recalling the defiition of ε it follows that

$$C_{2,k}(R) \le c^{k/\gamma_k}(s^p(R)(s_F(R) + 1) + t^p(R))^{1/\gamma_k}, \quad c > 1 .$$

Use Minkowski's inequality to get

$$(\frac{1}{\omega(B_R)} \int_{B_{k+1}} |v|^{\gamma_{k+1}}\omega)^{1/\gamma_{k+1}} \le cC_{2,k}(R)[(\frac{1}{\omega(B_R)} \int_{B_k} |v|^{\gamma_k})^{1/\gamma_k} + \gamma_k]$$

Iterate this to get

$$\left(\frac{1}{\omega(B_R)}\int |v|^{\gamma_{n+1}}\omega\right)^{1/\gamma_{n+1}} \leq \prod_{k=0}^{n} C_{2,k}(R)\left(\frac{1}{\omega(B_R)}\int_{B_{\rho_2}} |v|^p\omega\right)^{1/p}$$

$$+ \sum_{j=0}^{n} \left(\prod_{k=j}^{n} C_{2,k}(R)\right)\gamma_j .$$

$C_{2,k}(R) \geq 1$ so $\sum_{j=0}^{n} \left(\prod_{k=j}^{n} C_{2,k}(R)\right)\gamma_j \leq \sum_{j=0}^{n} \gamma_j \prod_{k=0}^{\infty} C_{2,k}(R) \leq$ $c\gamma_n \prod_{k=0}^{\infty} C_{2,k}(R) \leq c\gamma_n(s(R)(s_F^{1/p}(R) + 1) + t(R))^{q/(q-p)}$.

Let $C_2(R) = c(s(R)(s_F^{1/p}(R) + 1) + t(R))^{q/(q-p)}$, then

$$\left(\frac{1}{\omega(B_R)}\int_{B_{n+1}} |v|^{\gamma_{n+1}}\omega\right)^{1/\gamma_{n+1}} \leq C_2(R)\left[\left(\frac{1}{\omega(B_R)}\int_{B_{\rho_2}} |v|^p\omega\right)^{1/p} + \gamma_n\right] .$$

Given $s \geq p$, then $\gamma_n \leq s \leq \gamma_{n+1}$ for some $n \geq 0$. Use Hölder's inequality, $\omega(B_{n+1})/\omega(B_R) \leq 1$ and $B_{\rho_1} \subseteq B_{n+1}$ to get

$$\left(\frac{1}{\omega(B_R)}\int_{B_{\rho_1}} |v|^s\omega\right)^{1/s} \leq C_2(R)\left[\left(\frac{1}{\omega(B_R)}\int_{B_{\rho_2}} |v|^p\omega\right)^{1/p} + s\right] . \qquad (3.1.36)$$

Expanding e^x in a Taylor series, it follows that

$$\frac{1}{\omega(B_R)}\int_{B_{\rho_1}} e^{a|v|}\omega = \sum_{s=0}^{\infty} \frac{a^s}{s!}\left(\frac{1}{\omega(B_R)}\int_{B_{\rho_1}} |v|^s\omega\right) .$$

Use Hölder's inequality on the first [p] terms and (3.1.36) on the rest to get

$$\frac{1}{\omega(B_R)}\int_{B_{\rho_1}} e^{a|v|}\omega \leq \sum_{s=0}^{\infty} \frac{a^s}{s!} C_2^s(R)\left(\left(\frac{1}{\omega(B_R)}\int_{B_{\rho_2}} |v|^p\omega\right)^{1/p} + s\right)^s .$$

The series $\sum_{s=0}^{\infty} \frac{b^s}{s!}(x + s)^s$ converges and is bounded by ce^x if $b < e^{-1}$, so

$$\frac{1}{\omega(B_R)}\int_{B_{\rho_1}} e^{a|v|} \leq c \exp\left(\frac{1}{\omega(B_R)}\int_{B_{\rho_2}} |v|^p\omega\right)^{1/p}$$

$$\text{if } a < e^{-1}C_2^{-1}(R) \leq ce^{c(p(R)H(R)+q(R))}$$

since (3.1.33) holds. Finally,

$$\frac{1}{\omega(B_R)}\int_{B_{\rho_1}} u^{\pm a}\omega = \frac{1}{\omega(B_R)}\int_{B_{\rho_1}} e^{\pm a \log u}\omega$$

$$\leq ce^{c(p(R)H(R)+q(R))}k^{\pm a} ,$$

and so

$$\left(\frac{1}{\omega(B_R)}\int_{B_{\rho_1}} u^{a}\omega\right)^{1/a}\left(\frac{1}{\omega(B_R)}\int_{B_{\rho_1}} u^{-a}\omega\right)^{1/a} \tag{3.1.37}$$

$$\leq ce^{c(p(R)H(R)+q(R))/a}$$

if $a < e^{-1}C_2^{-1}(R)$.

<u>3.1.38.</u> Let σ from 3.1.30 be $e^{-1}C_2^{-1}(R)$ so that $\exists\, a$ such that $p\sigma/q < a < \sigma$ and (3.1.31) holds with $\gamma_0 = \pm a$. With this choice of γ_0 it follows that $|\gamma_0|s_F^{1/p}(R) \leq 1$ so that $C_1(R) \leq c(s(R) + t(R))^{pq/(|\gamma_0|(q-p))}$. Combining (3.1.31) and (3.1.37) it now follows that

$$\sup_{B_\rho} u \leq C(R) \inf_{B_\rho} u$$

with

$$C(R)=[(s(R)+ t(R))\exp(p(R)H(R) + q(R))]^{c(s(R)(s_F^{1/p}(R)+1)+t(R))^{q/(q-p)}}$$

since it was assumed that $s(R) \geq 2$ for simplicity. ∎

<u>Proof of Corollary 3.1.12.</u> It is easy to see that if u is a solution of (3.1.1), then $-u$ is a solution of an equation which is almost identical to (3.1.1) and which satisfies conditions (3.1.2). Because of this it is only necessary to deal with essential minima.

Since $a_1 = a_2 = b_2 = b_3 = c_1 = c_2 = 0$, it is seen that $V - c$ is a solution for any constant c. Assume V has an essential minimum at x_0 and let $u = V - \inf_{B(x_0,R)} V$, and apply Theorem 3.1.10 for any θ, $0 < \theta < 1$, to show that

$$\sup_{B(x_0,\theta R)} V - \inf_{B(x_0,R)} V \leq C(R)\left(\inf_{B(x_0,\theta R)} V - \inf_{B(x_0,R)} V\right) = 0 ,$$

and so $V = \inf_{B(x_0,R)} V = \inf_{\Omega} V$ a.e. in $B(x_0,\theta R)$. But then the set of points where V achieves its essential minimum is both open and closed so the connectedness of Ω implies that $V = \inf_{\Omega} V$ a.e. in Ω. ∎

Proof of Corollary 3.1.13. Let $\bar{u} = u + K(R)$ and define

$$\bar{A}(x,\bar{u},\nabla\bar{u}) = A(x,\bar{u} - K(R),\nabla\bar{u})$$

and

$$\bar{B}(x,\bar{u},\nabla\bar{u}) = B(x,\bar{u} - K(R),\nabla\bar{u}) ,$$

so that

$$|\bar{A}(x,\bar{u},\nabla\bar{u})| = |A(x,u,\nabla u)|$$

$$\leq \mu|\nabla u|^{p-1} + a_1 u^{p-1} + a_2$$

$$\leq \mu|\nabla\bar{u}|^{p-1} + (a_1 + a_2 K^{-(p-1)}(R))\bar{u}^{p-1} .$$

Similarly,

$$|\bar{B}(x,\bar{u},\nabla\bar{u})| \leq b_0\lambda|\nabla\bar{u}|^p + b_1|\nabla\bar{u}|^{p-1} + (b_2 + b_3 K^{-(p-1)}(R))\bar{u}^{p-1}$$

and

$$\bar{A}(x,\bar{u},\nabla\bar{u})\cdot\nabla\bar{u} \geq \lambda|\nabla\bar{u}|^p - (c_1 + c_2 K^{-p}(R))\bar{u}^p$$

and $\operatorname{div} \bar{A} = \bar{B}$.

Now apply Theorem 3.1.10 to $\bar{u}$ to get

$$\sup_{B(x_0,\theta R)} \bar{u} \leq C(R) \inf_{B(x_0,\theta R)} \bar{u} ,$$

and (3.1.14) follows.

Proof of Theorem 3.1.15. Let $\bar{u}_1 = u - \inf_{B_r} u$, $\bar{u}_2 = \sup_{B_r} u - u$, $B_r = B(x_0,r)$,

$$\bar{A}_1(x,\bar{u}_1,\nabla\bar{u}_1) = A(x,\bar{u}_1 + \inf_{B_r} u,\nabla\bar{u}_1) ,$$

$$\bar{A}_2(x,\bar{u}_2,\nabla\bar{u}_2) = A(x,\sup_{B_r} u - \bar{u}_2,-\nabla\bar{u}_2) ,$$

and similarly for $\bar{B}_i$, $i = 1,2$, so that $\bar{u}_i$, $i = 1,2$, are solutions to the equations $\operatorname{div} \bar{A}_i = \bar{B}_i$, $i = 1,2$, which satisfy the structure

$$|\bar{A}_i| \leq \mu|\nabla\bar{u}_i|^{p-1} + (a_1 M + a_2) ,$$

$$|\bar{B}_i| \leq b_0\lambda|\nabla\bar{u}_i|^p + b_1|\nabla\bar{u}_i|^{p-1} + (b_2 M + b_3)$$

$$\bar{A}_i\cdot\nabla\bar{u}_i \geq \lambda|\nabla\bar{u}_i|^p - (c_1 M + c_2) .$$

Now apply Corollary 3.1.13 to get that

$$\sup_{B_{r\theta}} \bar{u}_i \le C(r) \inf_{B_{r\theta}} \bar{u}_i + (C(r) - 1)K(r) \quad \text{for} \quad 0 < r < R .$$

Adding these inequalities gives

$$\operatorname{Osc}_{B_{r\theta}} u + \operatorname{Osc}_{B_r} u \le C(r)(\operatorname{Osc}_{B_r} u - \operatorname{Osc}_{B_{r\theta}} u) + 2(C(r) - 1)K(r) ,$$

and so

$$\operatorname{Osc}_{B_{r\theta}} u \le \frac{C(r) - 1}{C(r) + 1} (\operatorname{Osc}_{B_r} u + 2K(r)) .$$

Now apply Lemma 3.1.16 to finish. ∎

Proof of Lemma 3.1.16. Iterating (3.1.18) easily gives (3.1.19). Assume (3.1.20), so

$$\log \prod_{k=0}^{n-1} \frac{C(r_k) - 1}{C(r_k) + 1} = \sum_{k=0}^{n-1} \log\Bigl(1 - \frac{2}{C(r_k) + 1}\Bigr) \tag{3.1.39}$$

$$\sim -2 \sum_{k=0}^{n-1} \frac{1}{C(r_k) + 1} \sim - \sum_{k=0}^{n-1} \frac{1}{C(r_k)}$$

since $C(r_k) \ge 1$, so $\prod_{k=0}^{n-1} a_k \to 0$. Furthermore,

$$2 \sum_{j=0}^{n-1} K(r_j) \prod_{k=j}^{n-1} a_k = \sum_{j=N}^{n-1} K(r_j)(C(r_j) - 1)\Bigl(\prod_{k=j+1}^{n-1} a_k - \prod_{k=j}^{n-1} a_k\Bigr) \tag{3.1.40}$$

$$+ 2 \sum_{j=0}^{N-1} K(r_j) \prod_{k=j}^{n-1} a_k$$

$$\le \sup_{j \ge N} K(r_j)C(r_j) + 2\Bigl(\prod_{k=0}^{n-1} a_k\Bigr) \sum_{j=0}^{N-1} K(r_j)\Bigl(\prod_{k=0}^{j-1} a_k\Bigr)^{-1} ,$$

so given $\varepsilon > 0$, pick N such that $K(r_j)C(r_j) < \varepsilon$ for $j \ge N$. Then, since $\prod_{k=0}^{n-1} a_k \to 0$ as $n \to \infty$, it is possible to pick $n > N$ so that the second term is less than ε. Therefore (3.1.21) is proven.

To prove 3.1.22, first assume that $\sum_{k=0}^{\infty} \frac{1}{C(r_k)} < \infty$, so from (3.1.39) it follows that $\prod_{k=0}^{n-1} a_k \ge C > 0$, and so $\prod_{k=0}^{n-1} a_n$ converges to a strictly positive number since $a_k < 1$. It is possible to choose u such that

$$\operatorname*{Osc}_{B_{r_k}} u = a_k \operatorname*{Osc}_{B_{r_{k-1}}} u, \qquad k = 1,2,\ldots,$$

but then $\lim_{n\to\infty} \operatorname*{Osc}_{B_{r_n}} u = \prod_{k=0}^{\infty} a_k \operatorname*{Osc}_{B_r} u > 0$. The othe possibility is that that $C(r_k)K(r_k) > c > 0$, in which case pick u such that

$$\operatorname*{Osc}_{B_{r_k}} u = \varepsilon = \left(1 - \frac{1}{\delta}\right)c ,$$

so

$$\varepsilon < \left(1 - \frac{1}{\delta}\right)C(r_k)K(r_k) < (C(r_k) - 1)K(r_k)$$

by (3.1.17), and so

$$\varepsilon(1 - a_k) < 2a_k K(r_k) \quad \text{and} \quad \varepsilon < a_k(\varepsilon + 2K(r_k)) ,$$

satisfying (3.1.18).

To prove (3.1.23), pick $M, c, \alpha > 0$ such that $C(r_k) < M$, $K(r_k) < cr_k^{\alpha}$, $k = 0,1,2,\ldots$.

$$a_k = 1 - \frac{2}{C(r_k) + 1} < 1 - \frac{2}{M + 1} = a = \theta^b$$

for some $b > 0$, and so

$$\operatorname*{Osc}_{B_{r_n}} u < a^{n-1} \operatorname*{Osc}_{B_R} u + 2cR^{\alpha} \sum_{j=0}^{n-1} \theta^{\alpha j} a^{n-j}$$

$$< (\theta R)^{-b} r_n^b \operatorname*{Osc}_{B_R} u + 2cR^{\alpha}\theta^{nb} \sum_{j=0}^{n-1} \theta^{(\alpha-b)j} < c' r_n^{\alpha'}$$

for some c', $\alpha' > 0$, since

$$\sum_{j=0}^{n-1} \theta^{(\alpha - b)j} < \begin{cases} \dfrac{1}{1 - \theta^{\alpha-b}} & \alpha - b > 0 \\ n & \alpha = b \\ c''\theta^{(\alpha-b)n} & \alpha - b < 0 \end{cases}$$

To prove sharpness first assume that $C(r_k)$ is not bounded so that the monotonicity assumption gives $C(r_k) \uparrow \infty$. Now pick u such that

$$\operatorname*{Osc}_{B_{r_k}} u = a_{k-1} \operatorname*{Osc}_{B_{r_{k-1}}} u ,$$

so that

$$\operatorname*{Osc}_{B_{r_n}} u = \prod_{k=0}^{n-1} a_k \operatorname*{Osc}_{B_r} u .$$

If $\operatorname*{Osc}_{B_{r_n}} u < cr_n^\alpha$, then

$$\prod_{k=0}^{n-1} a_k < \frac{c\theta^{n\alpha}R^\alpha}{\operatorname*{Osc}_{B_R} u} .$$

Pick N such that $\theta^\alpha < \sigma < a_k < 1$ for $k > N$, so

$$\Big(\prod_{k=0}^{N} a_k\Big)\sigma^{n-N} < c'\theta^{n\alpha} ,$$

and then

$$\Big(\frac{\sigma}{\theta^\alpha}\Big)^n < c''$$

for $n > N$, which is impossible since $\delta > \theta^\alpha$. To show the necessity of $K(r_k) < cr^\alpha$, pick u such that (3.1.19) is satisfied with equality. Assume $\operatorname*{Osc}_{B_{r_n}} u < cr_n^\alpha$, $n = 0,1,\ldots,$ so that

$$\theta^{-\alpha}cr_{n-1}^\alpha = cr_n^\alpha > \operatorname*{Osc}_{B_{r_n}} u = \prod_{k=0}^{n-1} a_k \operatorname*{Osc}_{B_r} u + 2\sum_{j=0}^{n-1} K(r_j) \prod_{k=j}^{n-1} a_k$$

$$> 2a_{n-1}K(r_{n-1}) > 2\Big(1 - \frac{2}{\delta + 1}\Big)K(r_{n-1}) ,$$

and $K(r_k) < c'r_k^\alpha$, $k = 1,2,\ldots$.

Finally, assuming that $C(r)$ is nondecreasing and $K(r)$ is nonincreasing as $r \to 0$, it follows that

$$\prod_{k=j}^{n-1} a_k = e^{\sum_{k=j}^{n-1} \log a_k} < e^{-\sum_{k=j}^{n-1} 1/(C(r_k))} < e^{-c\sum_{k=j}^{n-1}\int_{r_{k+1}}^{r_k} (1/(C(r)))dr/r}$$

where $c = (\log \theta^{-1})^{-1}$, and so

$$\prod_{k=j}^{n-1} a_k < e^{-c(g(r_n)-g(r_j))} ,$$

where $g(r) = \int_r^R \frac{1}{C(r)}\frac{dr}{r}$.

Also

$$\sum_{j=0}^{n-1} K(r_j) \prod_{k=j}^{n-1} a_k \leq \sum_{j=0}^{n-1} K(r_j) e^{-c(g(r_n)-g(r_j))}$$

$$\leq c e^{-cg(r_n)} \sum_{j=0}^{n-1} \int_{r_{j+1}}^{r_1} K(\tfrac{s}{\theta}) e^{cg(s)} \frac{ds}{s}$$

$$= c e^{-g(r_n)} \int_{r_n}^{R} C(s) K(\tfrac{s}{\theta}) e^{cg(s)} |g'(s)| ds$$

$$= c e^{-g(r_n)} \int_0^{g(r_n)} \gamma \circ g^{-1}(t) e^{ct} dt \, ,$$

where $\gamma(s) = C(s)K(\frac{s}{\theta})$, so that

$$\operatorname*{Osc}_{B_{r_n}} u \leq e^{-cg(r_n)} (\operatorname*{Osc}_{B_R} u + 2 \int_0^{g(r_n)} \gamma \circ g^{-1}(t) e^{ct} dt) \, . \blacksquare$$

Boundary Continuity

In this section solutions which take on continuous boundary values in a weak sense will be shown to do so continuously. The definition of weak boundary values adopted is that introduced by Gariepy and Ziemer [GZ]. The local nature of this condition is more appropriate and less restrictive than the more usual global condition that $u - f \in W_0^{1,p}(\Omega)$ for boundary values $f \in W^{1,p}(\Omega')$, $\bar{\Omega} \subseteq \Omega'$ and f continuous in $\Omega' - \Omega$. In Proposition 2.2.86 it is shown that under certain circumstances the two are equivalent.

For $(u,\nabla u) \in W_{loc}^{1,p}(\Omega)$, $x_0 \in \partial\Omega$ and $\ell \in \mathbf{R}$, it is said that

$$u(x_0) \leq \ell \text{ weakly} \tag{3.1.41}$$

if for every $K > \ell$ there is an $r > 0$ such that $\eta(u - k)^+ \in W_0^{1,p}(\Omega)$ for all $\eta \in C_0^\infty(B(x_0,r))$. The condition

$$u(x_0) \geq \ell \tag{3.1.42}$$

is defined analogously and $u(x_0) = \ell$ weakly if both (3.1.41) and (3.1.42) hold.

Throughout this section it will be assumed that Ω, Ω' are open with $\bar{\Omega} \subseteq \Omega'$, and that μ, λ, a_i, C_i, $i = 1,2$, b_i, $i = 1,2,3$ are defined in Ω'.

Suppose that u is a bounded weak solution of (3.1.1) in Ω and that $x_0 \in \partial\Omega$ and $u(x_0) \leq \ell$ weakly. For $k > \ell$ let

$$u_k = \begin{cases} (u-k)^+, & \Omega \\ 0, & \Omega' - \Omega \end{cases}$$

so for some $R > 0$

$$\eta u_k \in W_0^{1,p}(\Omega) \quad \text{if} \quad \eta \in C_0^\infty(B(x_0,R)) \ . \tag{3.1.43}$$

In the setting of 3.1.48 it is shown that given $R > 0$, (3.1.43) holds for all $k > \sup_{B(x_0,r)\ \partial\Omega} f$. In any case, Theorem 3.1.44 holds for any R, k for which (3.1.43) holds. It will be assumed that the Sobolev inequalities (3.1.4), (3.1.5), (3.1.6) hold for $0 < r \leq R$, where

$$F(r) = r^p[(c_1 + c_2)K^{-p}(r) + b_1^p\lambda^{-(p-1)} + (b_2 + b_3)K^{-(p-1)}(r)$$
$$+(a_1 + a_2)^{p/(p-1)}K^{-p}(r)\mu^{-p/(p-1)}]$$

for some function $K(r)$, $0 < K(r) \leq 1$.

Let $\mu(r) = \sup_{B(x_0,r)} u_k$ and $\bar{u} = \mu(r) + K(r) - u_k$.

<u>3.1.44</u> <u>Theorem</u>. If u is a bounded weak solution of (3.1.1) in Ω with $|u| \leq M$, $x_0 \in \partial\Omega$ and $u(x_0) \leq \ell$ weakly, and $\eta \in C_0^\infty(B(x_0,r/2))$, with $|\nabla\eta| \leq \frac{\bar{c}}{r}$ and $r \leq R$, then

$$r^p \int |\nabla(\eta\bar{u})|^p\lambda$$
$$\leq (\mu(r) + K(r))(\mu(r) - \mu(\tfrac{r}{2}) + K(r))^{p-1}G(r)\omega(B(x_0,r)) \ ,$$

where

$$G(r) = [(s(r)+t(r))\exp(p(r)H(r)+q(r))]^{c(s(r)(s_F^{1/p}(r)+1)+t(r))^{q/(q-r)}}$$

and

$$H(r) = 1 + \frac{1}{\omega(B_r)}\int_{B_r} ([(c_1 + c_2)K^{-p}(r) + b_1^p\lambda^{-(p-1)}$$
$$+ (b_2 + b_3)K^{-(p-1)}(r)]r^p + (a_1 + a_2)K^{-(p-1)}(r)r^{p-1}) \ .$$

$B_r = B(x_0,r)$. c depends only on p, q, M, $\bar{c}$.

<u>Definition</u>. If $K \subseteq \Omega'$ is compact, then

$$C_\lambda(K) = \inf\{\int |\nabla\phi|^p\lambda : \phi \in C_0^\infty(\Omega'),\ \phi \geq 1 \ \text{on} \ K\} \ .$$

3.1.45 Theorem. Suppose u is a bounded solution of (3.1.1) in Ω and $|u| < M$, $x_0 \in \partial\Omega$ and $u(x_0) = \ell$ weakly. If

$$\int_0 \left(\frac{C_\lambda(\overline{B(x_0,r/4)} - \Omega)r^p}{\omega(B(x_0,r))G(r)}\right)^{1/(p-1)} \frac{dr}{r} = \infty \tag{3.1.46}$$

and $\int_0 K(r)\,\frac{dr}{r} < \infty$, then

$$\limsup_{\substack{x \to x_0 \\ x \in \Omega^0}} |u(x) - \ell| = 0 .$$

$G(r)$ is as in Theorem 3.1.44.

Remarks. Semicontinuity results for sub- and supersolutions and results on capacitary fine limits may be derived as in [GZ].

If u is taken to be quasicontinuous (see 2.2.20), as may be done if $C_\lambda(E) = 0 \Longrightarrow |E| = 0$ (a condition which follows from the existence of a Poincaré inequality), then

$$\sup u = \sup_c u , \tag{3.1.47}$$

where $\sup u = \inf\{t \in \mathbf{R} : u < t \text{ a.e.}\}$ and $\sup_c u = \inf\{t \in \mathbf{R} : u < t \text{ quasi-everywhere}\}$. Theorem 3.1.45 then states that $u(x) \to \ell$ as $x \to x_0$ for all $x \in \Omega$ outside of a set of capacity zero. (3.1.47) follows from the fact that u can be approximated pointwise quasi-everywhere by $\phi_n \in C^\infty$ such that $\phi_n < \sup u$. This is proven in Proposition 2.2.2.

Modulus of Continuity at the Boundary

3.1.48. Assume $f : \partial\Omega \to \mathbf{R}$ is continuous, and $u = f$ weakly on $\partial\Omega$. If $x_0 \in \partial\Omega$, let

$$V(r) = \sup_{B(x_0,r)\cap\Omega} (u - f(x_0))^+$$

and

$$M(r) = \sup_{B(x_0,r)\cap\partial\Omega} f - f(x_0) .$$

From (3.1.51) and (3.1.52) in the proof of Theorem 3.1.45 it follows that

$$A(r) = \left(\frac{C_\lambda(\overline{B(x_0,r/4)} - \Omega)r^p}{\omega(B(x_0,r)G(r)}\right)^{1/(p-1)} < \frac{\mu(r) - \mu(r/2) + K(r)}{\mu(r) + K(r)} < 1 ,$$

when r, k satisfy (3.1.43). But by Lemma 3.1.50 r, k satisfy

(3.1.43) if $k > \sup_{B(x_0,r)\cap\partial\Omega} f$. It is clear that $(u - f(x_0))^+ \le (u - k)^+ + k - f(x_0)$, so $V(r) \le \mu(r) + k - f(x_0)$. But $(u - k)^+ \le (u - f(x_0))^+$, so $\mu(r) - \mu(r/2) \le V(r) - V(r/2) + k - f(x_0)$, and letting $k \to \sup_{B(x_0,r)\cap\partial\Omega} f$ it is seen that

$$(V(r) - m(r) + K(r))A(r) \le (V(r) - V(r/2) + m(r) + K(r))$$

and

$$V(r/2) \le (1 - A(r))(V(r) + K(r)) + (1 + A(r))m(r)$$

$$\le \left(1 - \frac{A(r)}{2}\right)(V(r) + K(r) + 4m(r))$$

$$\le \left(\frac{C(r-1)}{C(r+1)}\right)(V(r) + 2\bar{K}(r))$$

for $C(r) = 4A^{-1}(r)$ and $\bar{K}(r) = \frac{K(r) + 4m(r)}{2}$. With this identification Lemma 3.1.16 applies. In particular:

3.1.49 Theorem. If $A(r) \ge c > 0$ and $K(r) + m(r) \le cr^{\alpha}$ for some $\alpha > 0$, then

$$\sup_{B(x_0,r)\cap\Omega} |u - f(x_0)| \le cr^{\alpha'}$$

for some $\alpha' > 0$.

Proof. By Lemma 3.1.16 and the calculations above, it follows that $V(r) \le cr^{\alpha'}$ for some $\alpha' > 0$. Do the calcualtions above for $-u$ and $-f$ to get $\sup_{B(x_0,R)\cap\Omega} (f(x_0) - u)^+ \le cr^{\alpha''}$, $\alpha'' > 0$, and the result follows. ∎

3.1.50 Lemma. If $u \in W^{1,p}_{loc}(\Omega)$, $f : \partial\Omega \to R$ is continuous with $u \le f$ weakly on $\partial\Omega$ and $x_0 \in \partial\Omega$, $r > 0$, then

$$\eta(u - k)^+ \in W_0^{1,p}(\Omega)$$

for $k > \sup_{B(x_0,r)\cap\partial\Omega} f$ for all $\eta \in C_0^\infty(B(x_0,r))$.

Proof. The result follows from a partition of unity argument. Assume $k > \sup_{B(x_0,r)\cap\partial\Omega} f$ so given $\bar{x} \in \partial\Omega \cap \overline{B(x_0,r)}$, $\exists \bar{r} > 0$ s.t. $\bar{r} < r$ and $\bar{\eta}(u - k)^+ \in W_0^{1,p}(\Omega)$ for $\bar{\eta} \in C_0^\infty(B(\bar{x},\bar{r}))$. The balls $B(\bar{x},\bar{r}/2)$ cover $\overline{B(x_0,r)} \cap \partial\Omega$, which is compact, so pick a finite subcover $B_i = B(x_i,r_i/2)$, $i = 1,2,\ldots,n$. Choose $\eta_i \in C_0^\infty(B(x_i,r_i))$ with $\eta_i = 1$ on B_i, and let $N = \overline{(B(x_0,r)} \cap \bar{\Omega}) - \bigcup_{i=1}^{n} B_i$. N is compact, and $N \subseteq \Omega$ since $\overline{B(x_0,r)} \cap \partial\Omega \subseteq \bigcup_{i=1}^{n} B_i$, so choose $\eta_0 \in C_0^\infty(\Omega)$ with $\eta_0 = 1$ on N. Since $\sum_{i=0}^{n} \eta_i \geq 1$ on $\bigcup_{i=1}^{n} B_i \cup N$, choose $\phi \in C_0^\infty(\{\sum_{i=0}^{n} \eta_i > \frac{1}{2}\})$ with $\phi = 1$ on $\bigcup_{i=1}^{n} B_i \cup N$. It follows that

$$\bar{\eta}_i = \frac{\eta_i \phi}{\sum_{i=0}^{n} \eta_i} \in C^\infty(\Omega')$$

and

$$\psi = \sum_{i=0}^{n} \bar{\eta}_i = 1$$

on $\bigcup_{i=1}^{n} B_i \cup N$. But $\bar{\eta}_i(u-k)^+ \in W_0^{1,p}(\Omega)$ since $\bar{\eta}_i \in C_0^\infty(B(x_i,r_i))$ for $i = 1,\dots,n$, and $\bar{\eta}_0 \in C_0^\infty(\Omega)$, so

$$\psi(u-k)^+ = \sum_{i=0}^{n} \bar{\eta}_i(u-k)^+ \in W_0^{1,p}(\Omega) .$$

But $\psi = 1$ on $B(x_0,r) \cap \Omega \subseteq \bigcup_{i=1}^{n} B_i \cup N$, so $\eta(u-k)^+ = \eta\psi(u-k)^+ \in W_0^{1,p}(\Omega)$ for all $\eta \in C_0^\infty(B(x_0,r))$, as required. ∎

Proof of Theorem 3.1.45. Given $k > \ell$, it is clear that $\bar{u} = \mu(r) + K(r)$ if $u_k = 0$. Take $\eta = 1$ on $B(x_0,r/4)$ so $\frac{\eta u}{\mu(r) + K(r)} = 1$ on $\bar{B}(x_0,r/4) \cap \{U_k = 0\}$, and

$$C_\lambda(\overline{B(x_0, \tfrac{r}{4}) \cap \{U_k = 0\}}) \tag{3.1.51}$$

$$\leq r^{-p}(\mu(r)+k(r))^{-(p-1)}(\mu(r)-\mu(\tfrac{r}{2})+k(r))^{p-1}\omega(B(x_0,r))G(r)$$

from (2.2.30), Theorem 3.1.44, and the definition of C_λ.

If $\Lambda = \limsup_{\substack{x\to x_0 \\ x\in\Omega}} u > \ell$, then for k such that $\ell < k < \Lambda$, it is clear that $\mu(r) \geq \Lambda - k > 0$. Also,

$$\overline{B(x_0,r)} - \Omega \subseteq \overline{B(x_0,r)} \cap \{u_k = 0\} \tag{3.1.52}$$

by the definition of u_k, so from (3.1.51) it follows that

$$\int_0 \left(\frac{C_\lambda(\overline{B(x_0,r/4)} - \Omega)r^p}{\omega(B(x_0,r)G(r)}\right)^{1/(p-1)} \frac{dr}{r}$$

$$\leq (\Lambda - k)^{-1} \int_0 (\mu(r) - \mu(r/2) + K(r)) \frac{dr}{r} .$$

This is finite since $\mu(r)$ is monotone increasing and $\int_0 K(r) \frac{dr}{r} < \infty$, but then (3.1.46) is contradicted, and so $\limsup_{\substack{x\to x_0 \\ x\in\Omega^0}} \leq \ell$

Since $-u$ is a solution of a slightly altered equation which satisfies the structure inequalities (3.1.2), it is also true that

$$\limsup_{\substack{x\to x_0 \\ x\in\Omega^0}} -u \leq -\ell ,$$

which completes the proof. ∎

Proof of Theorem 3.1.44. Let $\psi_\beta = u^{-\beta} - (\mu(r) + K(r))^\beta$ for $\beta \neq 0$. Given $0 < r < R$, R as in (3.1.43), assume $\phi \in C_0^\infty(B_r)$, $B_r = B(x_0,r)$, and choose $\eta \in C_0^\infty(B_R)$ so that $\eta = 1$ on B_r, $\eta u_k \in W_0^{1,p}(\Omega)$, so repeated applications of Proposition 2.2.2 show that $\Phi \in W_0^{1,p}(\Omega \cap B_r)$, where

$$\Phi = \phi^p \psi_\beta e^{b_0 \text{sign}\beta\, \bar{u}} = \phi^p((\mu(r) + k(r) - \eta u_k)^\beta$$

$$- (\mu(r) + k(r))^\beta)e^{b_0 \text{sign}\beta(\mu(r)+k(r)-\eta u_k)} ,$$

with gradient

$$\nabla\Phi = p\nabla\phi\phi^{p-1}\psi_\beta e^{b_0\text{sign}\beta\, \bar{u}} + \beta\phi^p\bar{u}^{-\beta-1}\nabla\bar{u}e^{b_0\text{sign}\beta\, \bar{u}}$$

$$+ \text{sign}\beta\, b_0\phi^p\psi_\beta e^{b_0\text{sign}\beta\, \bar{u}}\nabla\bar{u} .$$

Substitute this in (3.1.3), use $E = e^{b_0\text{sign}\beta\, \bar{u}}$, and multiply by $\text{sign}\beta$ to get

$$|\beta| \int \phi^p\bar{u}^{-\beta-1}\nabla\bar{u}\cdot AE$$

$$= -\text{sign}\beta \int \phi^p\psi_\beta BE - \text{sign}\beta\, p \int \phi^{p-1}\psi_\beta\nabla\phi\cdot AE - b_0 \int \phi^p\psi_\beta\nabla\bar{u}\cdot AE .$$

Use the structure inequalities (3.1.2) with the fact that ϕ, $\nabla\bar{u}$, and ψ_β are supported in Ω, $\psi_\beta = 0$ on $\{u < k\}$, and $\nabla\bar{u} = -\chi_{\{u>k\}}\nabla u$ to get

$$|\beta| \int \phi^p \bar{u}^{\beta-1} |\nabla \bar{u}|^p \lambda E \le |\beta| \int \phi^p \bar{u}^{\beta-1} (c_1|u|^p + c_2)E$$

$$+ b_0 \int \phi^p \psi_\beta |\nabla \bar{u}|^p \lambda E + \int \phi^p \psi_\beta |\nabla \bar{u}|^{p-1} b_1 E$$

$$+ \int \phi^p \psi_\beta (b_2|u|^{p-1} + b_3)E + p \int |\nabla\phi| \phi^{p-1} \psi_\beta |\nabla \bar{u}|^{p-1} \mu E$$

$$+ p \int |\nabla \phi| \phi^{p-1} \psi_\beta (a_1|u|^{p-1} + a_2)E$$

$$- b_0 \int \phi^p \psi_\beta |\nabla \bar{u}|^p \lambda E + b_0 \int \phi^p \psi_\beta (c_1|u|^p + c_2)E .$$

The second and the seventh terms cancel. Use

$$\psi_\beta \le u^{-\beta}, \quad |u| \le M, \quad 1 \le \bar{u}K^{-1}(r) , \tag{3.1.53}$$

$$\bar{a} = (a_1 + a_2)K^{-(p-1)}(r), \quad \bar{b} = (b_2 + b_3)K^{-(p-1)}(r) ,$$

$$\bar{c} = (c_1 + c_2)K^{-p}(r), \quad \text{and} \quad K(r) \le 1 ,$$

to get

$$|\beta| \int \phi^p \bar{u}^{\beta-1} |\nabla \bar{u}|^p \lambda E \le c(\beta + 1) \int \phi^p \bar{u}^{\beta+p-1} (\bar{b} + \bar{c})E \tag{3.1.54}$$

$$+ \int \phi^p \bar{u}^\beta |\nabla \bar{u}|^{p-1} b_1 E + p \int |\nabla\phi| \phi^{p-1} \bar{u}^\beta |\nabla \bar{u}|^{p-1} \mu E$$

$$c \int |\nabla \phi| \phi^{p-1} \bar{u}^{\beta+p-1} \bar{a} E .$$

Much of the rest of the proof follows that of Theorem 3.1.10. Continue as from (3.1.25) with minor changes such as the redefinition of $F(r)$ to get as in (3.1.27) and (3.1.29) that

$$r^p \int \phi^p |\nabla \bar{u}^{\gamma/p}|^p \lambda \le cC(\gamma) \int (\varepsilon^{-\delta} s_F(r) \phi^p + r^p |\nabla \phi|^p) \bar{u}^\gamma \omega \tag{3.1.55}$$

for $\gamma \ne p - 1$, $\gamma \ne 0$, and

$$\left(\frac{1}{\omega(B_r)} \int_{B_{k+1}} \bar{u}^{\gamma_{k+1}} \omega\right)^{1/|\gamma_{k+1}|} \le C_{1,k}(r) \left(\frac{1}{\omega(B_r)} \int_{B_k} u^{\gamma_k} \omega\right)^{1/|\gamma_k|} , \tag{3.1.56}$$

where

$C_{1,k}(r) =$

$$c\left(s^p(r)\left[c(\gamma_k)\left(\varepsilon_k^{-\delta} s_F(r) + \left(\frac{2^{k+2}}{\theta_1 - \theta_0}\right)^p\right) + \left(\frac{2^{k+2}}{\theta_1 - \theta_0}\right)^p\right] + t^p(r)\right)^{1/|\gamma_k|}$$

Given a s.t., $0 < a < p - 1$, iterate as in (3.1.31) with $\gamma_0 = -a < 0$, to get

$$\left(\frac{1}{\omega(B_r)}\int_{B_{\rho_1}}\bar{u}^{-a}\omega\right)^{-1/a} \leq C_1(r)\inf_{B_\rho}\bar{u}\ , \tag{3.1.57}$$

with $C_1(r) = c\left(s(r)(|\gamma_0|s_F^{1/p}(r) + 1) + t(r)\right)^{pq/|\gamma_0|(q-p)}$.

As in 3.1.38, γ_0 will be chosen so that

$$C_1(r) \leq c\left(s(r) + t(r)\right)^{pq/|\gamma_0|(q-p)} .$$

If $\gamma_0 = a > 0$, iterate inequality (3.1.56) only for $k \leq k_0$, where $\gamma k_0 < p - 1$. By choosing θ_1 and θ_2 at 3.1.28 slightly differently the iteration gives

$$\left(\frac{1}{\omega(B_r)}\int_{B_\rho}\bar{u}^{q\delta/p}\right)^{p/q\delta} \leq C_1(r)\left(\frac{1}{\omega(B_r)}\int_{B_{\rho_1}}\bar{u}^{a}\right)^{1/a} \tag{3.1.58}$$

for $\delta = \gamma_{k_0} < p - 1$.

The "crossover" also follows that of Theorem 3.1.10 closely. Proceeding as from 3.1.32, using (3.1.54) instead of (3.1.24) and letting $K_1 = \frac{1}{\omega(B_{\rho_2})}\int_{B_{\rho_2}}\log\bar{u}\,\omega$, it follows that

$$\rho_2^p\int_{B_{\rho_2}}\left|\log\frac{\bar{u}}{K_1}\right|^p\omega \leq c(p(r)H(r) + q(r)) ,$$

where

$$H(r) = 1 + \frac{1}{\omega(B_r)}\int_{B_r}\Big([(c_1 + c_2)K^{-p}(r) + b_1^p\lambda^{-(p-1)}$$

$$+ (b_2 + b_3)K^{-(p-1)}(r)]r^p + (a_1 + a_2)K^{-(p-1)}(r)r^{p-1}\Big) .$$

Now let $\phi = \eta^p\psi(|v|^\beta + (\frac{p\beta}{p-1})^\beta)e^{-b_0\bar{u}}$ for $\beta > 1$ with $\eta \in C_0^\infty(B_r)$, $\psi = \bar{u}^{1-p} - (\mu(r) + K(r))^{1-p}$, and $v = \log\frac{\bar{u}}{K_1}$, K_1 as above, and

$$\nabla\phi = p\eta^{p-1}\nabla\eta\psi(|v|^\beta + (\frac{p\beta}{p-1})^\beta)e^{-b_0\bar{u}}$$

$$- \eta^p\bar{u}^{-p}\nabla\bar{u}\left((p-1)(|v|^\beta + (\frac{p\beta}{p-1})^\beta) - \beta\psi\bar{u}^{p-1}|v|^{\beta-1}\operatorname{sign} v\right)e^{-b_0\bar{u}}$$

$$- b_0\eta^p\psi(|v|^\beta + (\frac{p\beta}{p-1})^\beta)e^{-b_0\bar{u}}\nabla\bar{u} .$$

Proceed as from 3.1.34 recalling that $\psi \leq \bar{u}^{1-p}$ and that η, $\nabla\bar{u}$, ψ are supported in Ω, then continue as in (3.1.53) to get

$$\int \phi^p \bar{u}^{-p} |\nabla \bar{u}|^p (|v|^\beta + (\frac{p\beta}{p-1})^\beta)\lambda \le c(\int \phi^p(\bar{b} + \bar{c}))(|v|^\beta + (\frac{p\beta}{p-1})^\beta)$$

$$+ \int \phi^p \bar{u}^{1-p}(|v|^\beta + (\frac{p\beta}{p-1})^\beta)|\nabla \bar{u}|^{p-1}\mu$$

$$+ \int \phi^{p-1}|\nabla \phi|(|v|^\beta + (\frac{p\beta}{p-1})^\beta)\bar{a}$$

$$+ \int \phi^p \bar{u}^{1-p}(|v|^\beta + (\frac{p\beta}{p-1})^\beta)|\nabla \bar{u}|^{p-1} b_1 .$$

Proceed as from 3.1.35 with minor changes as in the redefinition of $F(r)$ to get

$$(\frac{1}{\omega(B_r)} \int_{B_{\rho_1}} \bar{u}^a \omega)^{1/a} (\frac{1}{\omega(B_r)} \int_{B_{\rho_1}} u^{-a} \omega)^{1/a} \qquad (3.1.59)$$

$$\le ce^{c(p(r)H(r)+q(r))/a}$$

for $a < e^{-1}C_2^{-1}(r)$, $C_2(r) = c(s(r)(s_F^{1/p}(r) + 1) + t(r))^{q/(q-p)}$.

Given δ s.t. $\frac{(p-1)}{2}\frac{p}{q} < \delta < p - 1$, pick a and k_0 s.t. $a\frac{p}{q} < \min\{e^{-1}C_2^{-1}(r), \frac{(p-1)}{2}\frac{p}{q}\} < a$ and $\gamma_{k_0} = a(\frac{q}{p})^{k_0} = \delta$.

Now combine (3.1.57), (3.1.58), and (3.1.59) to get

$$(\frac{1}{\omega(B_r)} \int_{B_\rho} \bar{u}^{q\delta/p}\omega)^{p/q\delta} \le C_3(r) \inf_{B_\rho} \bar{u}$$

with $C_3(r) \le cC_1^2(r)e^{c(p(r)H(r)+q(r))/a}$.

Taking $\rho = \frac{r}{2}$ and recalling that $\bar{u} = \mu(r) - u_K + K(r)$, it follows that

$$(\frac{1}{\omega(B_r)} \int_{B_{r/2}} \bar{u}^{q\delta/p}\omega)^{p/q\delta} \le C_3(r)(\mu(r) - \mu(r/2) + K(r)) \qquad (3.1.60)$$

for $\frac{(p-1)}{2}\frac{q}{p} < \delta < p - 1$.

For the final step let $\Phi = \phi^p e^{-b_0\bar{u}} u_K$, with $\phi \in C_0^\infty(B_{r/2})$, $|\nabla\phi| \le \frac{c}{r}$ and $|\phi| \le c$.

$$\nabla\Phi = p\nabla\phi\phi^{p-1}e^{-b_0\bar{u}}u_K + \phi^p e^{-b_0\bar{u}}\nabla u_K$$

$$+ b_0\phi^p e^{-b_0\bar{u}} u_K \nabla u_K \quad (\nabla\bar{u} = -\nabla u_K) ,$$

and following the usual procedure it is seen that

$$\int \phi^p \nabla u_K \cdot AE = - \int \phi^p u_K BE - p\int \phi^{p-1} u_K \nabla\phi \cdot AE - b_0 \int \phi^p u_K \nabla u_K \cdot AE ,$$

and

$$\int \phi^p |\nabla u_k|^p \lambda E \le \int \phi^p (c_1 |u|^p + c_2) E + b_0 \int \phi^p u_k |\nabla u_k|^p \lambda E$$

$$+ \int \phi^p u_k |\nabla u_k|^{p-1} b_1 E + \int \phi^p u_k (b_2 |u|^{p-1} + b_3) E$$

$$+ p \int \phi^{p-1} |\nabla \phi| u_k |\nabla u_k|^{p-1} \mu E$$

$$+ p \int \phi^{p-1} |\nabla \phi| u_k (a_1 |u|^{p-1} + a_2) E$$

$$- b_0 \int \phi^p u_k |\nabla u_k|^p \lambda E + b_0 \int \phi^p u_k (c_1 |u|^p + c_2) E .$$

Cancel the second and seventh terms and use an inequality similar to (3.1.26), $|u| \le M$, and $K(r) \le 1$ to get

$$r^p \int \phi^p |\nabla u_k|^p \lambda \le c(\mu(r) + K(r)) K^{p-1}(r) \int (\phi^p r^p (c_1 + c_2) K^{-p}(r)$$

$$+ (b_2 + b_3) K^{-(p-1)}(r) + (a_1 + a_2)^{p/(p-1)} \mu^{-p/(p-1)} \lambda K^{-p}(r)$$

$$+ r^p |\nabla \phi|^p \omega) + r^p \int \phi^{p-1} u_k |\nabla u_k|^{p-1} (\phi b_1 + |\nabla \phi| \mu) .$$

Use inequality (3.1.6) with $\varepsilon = 1$ and $\Phi = \phi$ in the first expression along with $\nabla u_k = -\nabla \bar{u}$ to get

$$r^p \int \phi^p |\nabla \bar{u}|^p \lambda \le c(s_F(r) + 1)(\mu(r) + K(r)) K^{p-1}(r) \omega(B_r) \qquad (3.1.61)$$

$$+ r^p \int \phi^{p-1} u_k |\nabla \bar{u}|^{p-1} (\phi b_1 + |\nabla \phi| \mu) .$$

Pick $\alpha > \frac{p-1}{2p}$ so that $1 < (1-\alpha)p < \frac{q}{p}$. Use inequality (3.1.6) with $\varepsilon = 1$, $\Phi = \phi \bar{u}^{\gamma^*/p}$, $\gamma^* = p(1-\alpha)(p-1)$ to get

$$r^p \int \phi^p \bar{u}^{\gamma^*} b_1^p \lambda^{-(p-1)} \le r^p \int \phi^p F(r) \qquad (3.1.62)$$

$$\le r^p \int \phi^p |\nabla \bar{u}^{\gamma^*/p}|^p \lambda + c(s_F(r) + 1) \int_{B_{r/2}} \bar{u}^{\gamma^*} \omega$$

$$\le c(s_F(r) + 1) \int_{B_{r/2}} \bar{u}^{\gamma^*} \omega \quad \text{by using 3.1.55.}$$

Also

$$r^p \int \phi^{p-1} u_k |\nabla \bar{u}|^{p-1} (\phi b_1 + |\nabla \phi| \mu)$$

$$\le \mu(r) \int (\phi^{p-1} \bar{u}^{(\alpha-1)(p-1)} r^{p-1} |\nabla \bar{u}|^{p-1} \lambda^{(p-1)/p} \times$$

$$(\bar{u}^{(1-\alpha)(p-1)} r(\phi b_1 + |\nabla\phi|\mu)\lambda^{-(p-1)/p})$$

$$\leq c\mu(r)(r^p \int \phi^p |\nabla\bar{u}^\alpha|^p \lambda)^{(p-1)/p} \times$$

$$(r^p \int \bar{u}^{\gamma *}(\phi^p b_1^p \lambda^{-(p-1)} + |\nabla\phi|^p \omega))^{1/p}$$

$$\leq c(s_F(r) + 1)\mu(r)(\int_{B_{r/2}} \bar{u}^{\alpha p}\omega)^{(p-1)/p}(\int_{B_{r/2}} \bar{u}^{\gamma *}\omega)^{1/p}$$

using (3.1.55), (3.1.62)

$$\leq c(s_F(r) + 1)C_3^{p-1}(r)\mu(r)(\mu(r) - \mu(r/2) + K(r))^{p-1}\omega(B_r)$$

by using (3.1.60).

Combining this with (3.1.61) gives

$$r^p \int \phi^p |\nabla\bar{u}|^p \lambda$$

$$\leq c(s_F(r)+1)C_3^{p-1}(r)(\mu(r) + K(r))(\mu(r) - \mu(r/2) + K(r))\omega(B_r) .$$

Using (3.1.60) once more gives

$$r^p \int |\nabla(\phi\bar{u})|^p \lambda$$

$$\leq G(r)(\mu(r) + K(r))(\mu(r) - \mu(r/2) + K(r))^{p-1}\omega(B_r) ,$$

where

$$G(r) \leq [(s(r)+t(r))\exp(p(r)H(r)+q(r))]^{c(s(r)(s_F^{1/p}(r)+1)+t(r))^{q/(q-p)}}$$

using the assumption in (3.1.6) that $s(r) \geq 2$.

3.2.0 Modulus of Continuity Estimates for Weighted Sobolev and V.M.O. Functions

A result of Morrey implies that functions in the unweighted Sobolev space $W^{1,p}(\mathbf{R}^d)$ are Hölder continuous for $p > d$. In the present section a similar result is proven for the weighted spaces $W^{1,p}(\omega,\nu,\Omega)$ which support a weak type of Sovolev inequality. This result is derived from an estimate for the modulus of continuity of functions of vanishing mean oscillation which in the unweighted case, for $g(x,r) = cr$, is due independently to Campanato [CA] and Meyers [MY2]. It will be used in Section 3.3.0 to establish continuity for solutions of certain degenerate elliptic systems

Functions of Vanishing Mean Oscillation

Let $g : R^d \times R^+ \to R^+$ be Borel measurable and let ω be a locally finite positive Borel measure on an open set $\Omega \subseteq \mathbf{R}^d$. For simplicity it will be assumed that

$$\omega(\partial B) = 0 \tag{3.2.1}$$

for all balls B with $\bar{B} \subseteq \Omega$. The theory is much more technical without this assumption.

MO_g will be the space consisting of functions $u : \mathbf{R}^d \to \mathbf{R}$, $u \in L^1_{loc}(\omega,\Omega)$ such that

$$\frac{1}{\omega(B)} \int_B \left| u(y) - \frac{1}{\omega(B)} \int_B u(z)d\omega(z)\right| d\omega(y) < g(x,r) \tag{3.2.2}$$

for all $B = B(x,r)$, $\bar{B} \subseteq \Omega$. The methods used allow for much more general sets than balls. This restrictive approach was chosen for simplicity.

To estimate $|u(y) - u(z)|$ for y, $z \in B$, $\bar{B} \subseteq \Omega$, it is only necessary to estimate $O(x) = |u(x) - \frac{1}{\omega(B)} \int_B u d\omega|$ for $x \in B$ since $|u(y) - u(z)| < O(y) + O(z)$.

To do this it is necessary to introduce some geometry.

<u>3.2.3.</u> Assume that $x \in B = B(x_0,R)$, and that $F_x : B \to [0,R)$ is continuous with $F_x^{-1}(0) = x$, $F_x^{-1}([0,R)) = B$, and $F_x^{-1}[0,r) = B_x(r)$ for $0 < r < R$, where $B_x(r)$ is a ball of radius r centered at $c_x(r) \in \mathbf{R}^d$. The continuity of F_x forces $c_x(r)$ to be continuous in r. Also, assume that $|x - c_x(r)| < \theta r$ for some θ, $0 < \theta < 1$, to insure that $\lim_{r\to 0} \frac{1}{\omega(B_x(r))} \int_{B_x(r)} u d\omega = u(x)$ a.e. ω. This follows from Propositions 1.1.3 and 1.1.5 and a covering result of A. P. Morse, page 6 [G]. Points where the limit above exists will be referred to as Lebesgue points.

Let $\bar{\omega}_x(r) = \omega(F_x^{-1}([0,r))) = \omega(B_x(r)) < \infty$ so that $\bar{\omega}_x(r)$ is a monotone increasing left-continuous function on $[0,R]$ and so induces a finite Borel measure $\underline{\omega}_x$ on $[0,R]$. $\bar{\omega}_x(r)$ is actually continuous in r because of (3.2.1), so $\underline{\omega}_x$ has no atoms. If f, E are Borel measurable and $f > 0$, then

$$\int_E f(r)\, d\underline{\omega}_x = \int_{F_x^{-1}(E)} f(F_x(y))d\omega(y) \,. \tag{3.2.4}$$

The method of proof of Theorem 3.2.5 involves a reduction to one dimension, where an integration by parts with respect to $\underline{\omega}_x$ is carried out. (3.2.4) indicates the flavor of the reduction, the basic difference being that $f(F_x(y))$ must be repalced by an arbitrary function $u(y)$.

<u>3.2.5 Theorem</u>. If $u \in MO_g$ and x, y are Lebesgue points of u with respect to ω, then

$$O(x) = |u(x) - \frac{1}{\omega(B)} \int_B u\,d\omega| \tag{3.2.6}$$

$$< 2g(x_0,R) + 4 \int_0^R \frac{g(c_x(r),r)}{\bar{\omega}_x(r)}\, d\underline{\omega}_x(r)$$

and so

$$|u(x) - u(y)| < 4\Big(g(x_0,R) + \int_0^R \frac{g(c_x(r),r)}{\bar{\omega}_x(r)}\, d\underline{\omega}_x(r) \tag{3.2.7}$$

$$+ \int_0^R \frac{g(c_y(r),r)}{\bar{\omega}_y(r)}\, \underline{\omega}_y(r)\Big) .$$

An alternate expression for (3.2.6), (3.2.7) results from (3.2.4) since

$$\int_0^R \frac{g(c_x(r),r)}{\bar{\omega}_x(r)}\, d\underline{\omega}_x(r) = \int_B \frac{g(c_x(F_x(z)),F_x(z))}{\omega(B_x(F_x(z)))}\, d\omega(z) .$$

3.2.8. If, in addition,

$$g(x,r) = f(\omega(B(x,r)))$$

for $f : \mathbf{R}^+ \to \mathbf{R}^+$ continuous, then

$$|u(x) - u(y)| < 8\Big(f(\omega(B)) + \int_0^{\omega(B)} \frac{f(s)}{s}\, ds\Big) .$$

Remarks

It should be noted that the assumption that x, y are Lebesgue points is usually unnecessary. In fact if $\int_0^R \frac{g(c_x(r),r)}{\bar{\omega}_x(r)}\, d\underline{\omega}(r) < \infty$ then it can be seen from the proof of Theorem 3.2.5 that the averages $\omega^{-1}(B_x(r)) \int_{B_x(r)} u\,d\omega$ from a Cauchy sequence in r and therefore converge. If the above integral is finite for all x in some set E then u may be redefined almost everywhere $-\omega$ in E so that every point is a Lebesgue point.

A typical geometry would be given by defining F_x implicitly as $|y - c_x(F_x(y))| = F_x(y)$, with $c_x(r) = x - \frac{r}{R}(x - x_0)$, $B = B(x_0,R)$, $\frac{|x - x_0|}{R} < 1$. $F(y)$ is the positive solution to a quadratic and its graph is a skewed cone with vertex x.

The geometry described at 3.2.3, with x generally off centered in B, is not necessary if ω is a doubling measure. That is, if

$\omega(B(x,2r)) \leq c\omega(B(x,r))$. In this case take $x = x_0 = c_x(r)$ and $F_x(y) = |x_0 - y|$ so that for x, y with $|x - y| = R$,

$$|u(x) - u(y)| \leq |u(x) - \frac{1}{\omega(B(x,R))} \int_{B(x,R)} u d\omega| \tag{3.2.9}$$

$$+ |u(y) - \frac{1}{(B(y,R))} \int_{B(y,R)} u d\omega| + E ,$$

where

$$E = \frac{1}{\omega(B(x,R))\omega(B(y,R))} \int_{B(x,R)} \int_{B(y,R)} |u(x) - u(y)| d\omega(x) d\omega(y)$$

$$\leq \frac{c}{\omega^2(B)} \int_B \int_B |u(x) - u(y)| d\omega(x) d\omega(y)$$

for $B = B(\frac{x + y}{2}, \frac{3}{2} R)$, using the doubling condition. Considering (3.2.10), this can be bounded using (3.2.2).

The continuity of f in (3.2.8) is not necessary. If $\underline{\omega}(\bar{\omega}^{-1}(B)) = 0$, where B is the set of non-Lebesgue points of f, then (3.2.8) still holds.

Proof of Theorem 3.2.5. (3.2.6) will be proven first, then (3.2.7) will follow immediately.

Assume that $u \in MO_g$ and that x is a Lebesgue point. $\omega(B_x(r)) > 0$ for all $r > 0$ since $\lim_{r\to 0} \frac{1}{\omega(B_x(r))} \int_{B_x(r)} u d\omega$ must be defined. For convenience the subscript x will be dropped from $\bar{\omega}_x$ and $\underline{\omega}_x$, and B_r will be the ball $B_r(x)$. $\int_B |u| d\omega < \infty$ since $u \in L^1_{loc}(\omega,\Omega)$ and $\bar{B} \subseteq \Omega$. Let $f(r) = \int_{B_r} u d\omega$. f is absolutely continuous with respect to ω (see 1.1.10) since $\forall\, \varepsilon > 0$, $\exists \delta > 0$ such that if $\omega(A) < \delta$, $A \subseteq B$, then $\int_A |u| d\omega < \varepsilon$, so that given disjoint intervals $I_i = [a_i,b_i)$, $i = 1,2,3,\ldots,$ with $\underline{\omega}(\bigcup_{i=1}^{\infty} I_i) < \delta$, then $\omega(A) = \underline{\omega}(\bigcup_{i=1}^{\infty} I_i) < \delta$, where $A = \bigcup_{i=1}^{\infty} (B_{b_i} - B_{a_i})$, so

$$\sum_{i=1}^{\infty} |f(b_i) - f(a_i)| \leq \int_A |u| d\omega < \varepsilon .$$

From Proposition 1.1.13 it follows that $f(r) = \int_{[0,r)} \frac{df}{d\underline{\omega}}(s) d\omega(s)$.

Denote $\frac{df}{d\underline{\omega}}$ by $\bar{u}$. It is easy to see for $u,v \in L^1_{loc}(\omega,\Omega)$ and $\lambda \in R$ that $\overline{u + \lambda v} = \bar{u} + \lambda \bar{b}$ a.e. ω.

It is claimed that $|\bar{u}| < \overline{|u|}$. Pick sets F_+, F_- disjoint so that $\bar{u}^+$, the positive part of $\bar{u}$, is supported on F_+; and $\bar{u}^-$, the negative part of $\bar{u}$, is supported on F_-, so

$$\int_E \bar{u}^+ d\underline{\omega} = \int_{E\cap F_+} \bar{u}\, d\underline{\omega} = \int_{\bar{\omega}^{-1}(E\cap F_+)} u\, d\omega$$

$$< \int_{\omega^{-1}(E\cap F)} u^+ d\omega = \int_{E\cap F} \overline{u^+}\, d\underline{\omega}$$

for all Borel measurable E. Therefore $\bar{u}^+ < \overline{u^+}$ and so $|\bar{u}| = \bar{u}^+ + (-\bar{u})^+ < \overline{u^+ + (-u)^+} = \overline{|u|}$.

From this it can be seen that

$$\int_0^r \int_0^r |\bar{u}(s) - \bar{u}(t)|\, d\underline{\omega}(s)\, d\underline{\omega}(t)$$

$$< \int_0^r \int_0^r \overline{|u(x) - \bar{u}(t)|}^{s} d\underline{\omega}(s)\, d\underline{\omega}(t)$$

$$= \int_0^r \int_{B_r} |u(x) - \bar{u}(t)|\, d\omega(x)\, d\underline{\omega}(t)$$

$$= \int_{B_r} \int_0^r |u(x) - \bar{u}(t)|\, d\underline{\omega}(t)\, d\omega(x)$$

$$< \int_{B_r} \int_{B_r} |u(x) - u(y)|\, d\omega(y)\, d\omega(x) \tag{3.2.10}$$

$$< \int_{B_r} \int_{B_r} \left|u(x) - \frac{1}{\omega(B_r)} \int_{B_r} u(z)\, d\omega(z)\right| d\omega(y)\, d\omega(x)$$

$$+ \int_{B_r} \int_{B_r} \left|\frac{1}{\omega(B_r)} \int_{B_r} u(z)\, d\omega(z) - u(y)\right| d\omega(y)\, d\omega(x)$$

$$< 2\omega^2(B_r)\, g(c(r),r) \quad \text{by (3.2.2)},$$

so $\frac{1}{\bar{\omega}^2(r)} \int_0^r \int_0^r |\bar{u}(s) - \bar{u}(t)|\, d\underline{\omega}(s)\, d\underline{\omega}(t) < 2g(c(r),r)$. Now let $A(r) = \frac{1}{\bar{\omega}(r)} \int_0^r \bar{u}(s)\, ds$, which is absolutely continuous with respect to $\underline{\omega}$ on $[\varepsilon, R)$, $\varepsilon > 0$ by Proposition 1.1.13 since $\bar{\omega}(r) > 0$ for $r > 0$. Also by differentiating $\bar{\omega}(r)A(r) = \int_0^r \bar{u}(s)\, ds$ with respect to

$\underline{\omega}$, as in 1.1.18, it is seen that $A(r) + \bar{\omega}(r)\,\frac{dA}{d\omega}(r) = \bar{u}(r)$ for $r > 0$, and so $\frac{dA}{d\underline{\omega}}(r) = \frac{\bar{u}(r) - A(r)}{\bar{\omega}(r)}$. Proposition 1.1.13 is used several times in the following.

$$|A(r) - A(\varepsilon)| = \left|\int_\varepsilon^r \frac{\bar{u}(s) - A(s)}{\bar{\omega}(s)}\, d\underline{\omega}(s)\right|$$

$$= \left|\int_\varepsilon^r \frac{1}{\bar{\omega}^2(s)} \int_0^s (\bar{u}(s) - \bar{u}(t))\, d\underline{\omega}(t)\, d\underline{\omega}(s)\right|$$

$$\leq \int_\varepsilon^r \frac{1}{\bar{\omega}^2(s)} \int_0^s |\bar{u}(s) - \bar{u}(t)|\, d\underline{\omega}(t)\, d\underline{\omega}(s)$$

$$= \frac{1}{\bar{\omega}^2(\rho)} \int_0^\rho \int_0^s |\bar{u}(s) - \bar{u}(t)|\, d\underline{\omega}(t)\, d\underline{\omega}(s) \Big|_{\rho=\varepsilon}^{\rho=r}$$

$$+ 2 \int_\varepsilon^r \frac{1}{\bar{\omega}^3(s)} \left(\int_0^s \int_0^\rho |\bar{u}(\rho) - \bar{u}(t)|\, d\underline{\omega}(t)\, d\underline{\omega}(\rho)\right) d\underline{\omega}(s)$$

$$\leq 2g(c(r),r) + 4 \int_\varepsilon^r \frac{g(c(s),s)}{\bar{\omega}(s)}\, d\underline{\omega}(s) ,$$

and the result follows by letting $\varepsilon \to 0$ since

$$A(\varepsilon) = \frac{1}{\bar{\omega}(\varepsilon)} \int_0^\varepsilon \bar{u}\, d\underline{\omega} = \frac{1}{\omega(B_\varepsilon)} \int_{B_\varepsilon} u\, d\omega \to u(x)$$

since x is assumed to be a Lebesgue point of u with respect to ω.

To prove 3.2.8, let

$$G(r) = \int_{\bar{\omega}(\varepsilon)}^{\bar{\omega}(r)} \frac{f(s)}{s}\, ds ,$$

G is absolutely continuous with respect to $\underline{\omega}$ since

$$\sum_i |G(b_i) - G(a_i)| = \sum_i \int_{\bar{\omega}(a_i)}^{\bar{\omega}(b_i)} \frac{f(s)}{s}\, ds \leq M_\varepsilon \sum_i \bar{\omega}[a_i, b_i) .$$

Differentiate using (1.1.33) to get $\frac{dG}{d\underline{\omega}} = \frac{f(\bar{\omega}(r))}{\bar{\omega}(r)}$, so that $G(r) = \int_\varepsilon^r \frac{f(\bar{\omega}(s))}{\bar{\omega}(s)}\, d\omega(s)$. Let $\varepsilon \to 0$ to finish the proof. ∎

Continuity of Sobolev Functions

Let $\Omega \subseteq R^d$ be an open set and let ω, ν be locally finite positive Borel measures on Ω with ν absolutely continuous to ω. Let $W^{1,p}(\omega,\nu,\Omega)$ be the Sobolev space defined in 2.2.1 with $p \geq 1$.

3.2.11 Theorem. Assume that $u \in W^{1,p}(\omega,\nu,\Omega)$ and that the Sovolev inequality

$$\frac{1}{\omega(B)} \int_B |u(x) - \frac{1}{\omega(B)} \int_B u(y)d\omega(y)|d\omega(x) \tag{3.2.12}$$

$$\leq K(x,r)(\int_B |\nabla u|^p d\nu)^{1/p}$$

holds for every $B = B(x,r)$ with $\bar{B} \subseteq \Omega$. Then, assuming the structure of Theorem 3.2.5,

$$|u(x) - u(y)| \leq 4\Big(K(x_0,R) + \int_0^R \frac{K(c_x(r),r)}{\bar{\omega}_x(r)} d\underline{\omega}_x(r)$$

$$+ \int_0^R \frac{K(c_y(r),r)}{\bar{\omega}_x(r)} d\underline{\omega}_y(r)\Big)\Big(\int_{B(x_0,R)} |\nabla u|^p d\nu\Big)^{1/p}$$

if x, y are Lebesgue points of u with respect to ω.

If, in addition, $K(x,r) = f(\omega(B(x,r)))$ for $f : R^+ \to R^+$ continuous, then

$$|u(x) - u(y)| \leq 8\Big(f(\omega(B(x_0,R))) + \int_0^{\omega(B(x_0,R))} \frac{f(s)}{s} ds\Big) \tag{3.2.13}$$

$$\cdot\Big(\int_{B(x_0,R)} |\nabla u|^p d\nu\Big)^{1/p} .$$

Proof. Apply Theorem 3.2.5 with

$$g(x,r) = K(x,r)\Big(\int_{B(x_0,R)} |\nabla u|^p d\nu\Big)^{1/p} . \quad \blacksquare$$

Remarks. Theorem 3.2.11 is true in the more general setting of Section 2.1.0.

By applying Hölder's inequality, it is seen that the inequality

$$\frac{1}{\omega(B)} \int_B |\phi(x) - \frac{1}{\omega(B)} \int_B \phi(y)d\omega(y)|^p d\omega(x) \tag{3.2.14}$$

$$\leq K^p(x,r) \int_B |\nabla\phi|^p d\nu$$

for all $B = B(x,r)$ with $\bar{B} \subseteq \Omega$ and all $\phi \in W^{1,p}(\omega,\nu,\Omega)$ is sufficient for (3.2.12) to hold. A limiting argument shows that (3.2.14) need only be assumed for $\phi \in C^{\infty}(\Omega)$. In Theorem 2.2.41 it is seen that (3.2.14) is equivalent to

$$\frac{\omega(K \cap B)}{\omega(B)} < cK(x,r)\bar{C}_H(K \cap B) \tag{3.2.15}$$

for compact sets $K \subseteq \Omega$ (recall that $\bar{B} \subseteq \Omega$ so $C_0^{\infty}(\Omega)|_B = C^{\infty}(\Omega)|_B$).

Example. If K is as in the example preceeding Lemma 3.1.16 and $\omega(x) = \nu(x)\text{dist}^{\alpha}(x,K)$ with $\alpha > -\gamma$, then (3.2.13) holds with $K(x,r) = cr\omega^{-1/p}(B(x,r))$ by Theorem 2.2.56 (specifically (2.2.62) and (2.2.63)). As in (2.2.70) it is seen that

$$\omega(B(x,r)) \sim r^d \max{}^{\alpha}\{r,\text{dist}(x,K)\} .$$

If $\alpha > 0$, then $r^{d+\alpha} < c\omega(B(x,r))$, so $K(x,r) = cr\omega^{-1/p}(B(x,r)) < f(\omega(B(x,r)))$ for $f(t) = ct^{\beta}$, $\beta = \frac{1}{d+\alpha} - \frac{1}{p}$. If $p > d + \alpha$, apply (3.2.13) to get

$$\begin{aligned}|u(x) - u(y)| &< c\omega^{\beta}(B(x_0,R)) \\ &< cR^{\beta d}\max{}^{\alpha\beta}\{R,\text{dist}(x_0,K)\} ,\end{aligned}$$

so u is Hölder continuous.

If $\alpha < 0$, then

$$r^d < \omega(B(x,r))\max{}^{|\alpha|}\{r,\text{dist}(x,K)\} < c\omega(B(x,r))$$

for bounded r, x, so a similar argument shows that u is Hölder continuous if $p > d$.

It is interesting to note that there is some additional regularity at $\bar{K}$ in the case of $\alpha < 0$. A careful analysis of the proof of Theorem 3.2.11 shows that

$$|u(x) - \frac{1}{\omega(B)} \int_B u d\omega| < cr^{1-(d+\alpha)/p}, \quad B = B(x,r) ,$$

if $p > d + \alpha$ and $x \in \bar{K}$, where $u(x)$ has been redefined on $\bar{K}$ in the natural way, that is, $u(x)$ is defined as the limit of the averages $\frac{1}{\omega(B(x,r))} \int_{B(x,r)} u d\omega$.

3.3.0 Higher Integrability for the Gradient of Solutions of Elliptic Systems with Application to Continuity of Solutions

Solutions u of second order quasilinear degenerate elliptic systems are considered. It is assumed that $u \in W^{1,p}_{loc}(\mu,\Omega)$, where p is the natural exponent for the equation. Then it is shown that $|\nabla u| \in L^{p+\varepsilon}_{loc}(\mu,\Omega)$. In the case of Lebesgue measure with $\mu \equiv 1$, this can be used to show that u is Hölder continuous in the borderline case $d - \varepsilon < p \leq d$. In the weighted case the critical exponent may change but continuity is still achieved. The analysis also applies to higher order equations. The basic method is due to N. Meyers and A. Elcrat [MYE] $(\mu = 1)$.

The equations considered are:

$$\nabla \cdot A_i(x,u,\nabla u) = B_i(x,u,\nabla u), \quad i = 1,\ldots,N , \tag{3.3.1}$$

where

$$A : \Omega \times \mathbf{R}^N \times \mathbf{R}^{dN} \to \mathbf{R}^d ,$$

and

$$B : \Omega \times \mathbf{R}^N \times \mathbf{R}^{dN} \to \mathbf{R}$$

are Borel measurable functions satisfying the conditions:

$$\begin{cases} \sum_i |A_i(x,u,\nabla u)| \leq a_0\mu|\nabla u|^{p-1} + a_1\mu , \\ \sum_i |B_i(x,u,\nabla u)| \leq b_1\mu|\nabla u|^{p-1} + b_2\mu , \\ \sum_i A_i \cdot \nabla i_i \geq \mu|\nabla u|^p - c_1\mu , \end{cases} \tag{3.3.2}$$

where $u = (u_1,\ldots,u_N)$, $\nabla u = (\nabla u_1,\ldots,\nabla u_N)$, $a_0 \geq 1$ is a constant, and μ, a_1, b_1, b_2, c_1 are nonnegative Borel measurable functions on the open set Ω. It is assumed that $0 < \mu < \infty$ a.e. and that μ induces a locally finite measure ω which is doubling, that is, $\omega(E) = \int_E \mu$ and $\omega(B(x,2r)) \leq c_\omega \omega(B(x,r))$. It is also assumed that there is an $\alpha > 0$ such that

$$a_1 \in L^{(p/(p-1))+\alpha}_{loc}(\mu,\Omega), \qquad b_1 \in L^{ps/(s-p)+\alpha}_{loc}(\mu,\Omega) , \tag{3.3.3}$$

$$b_2 \in L^{s/(s-1)+\alpha}_{loc}(\mu,\Omega), \quad \text{and} \quad c_1 \in L^{1+\alpha}_{loc}(\mu,\Omega) ,$$

where s is given in (3.3.4), (3.3.5).

In some cases these conditions can be weakened by using other Sobolev inequalities in combinations with those in (3.3.4), (3.3.5).

Let $W^{1,p}_{loc}(\mu,\Omega) = \prod_{i=1}^{N} W^{1,p}_{loc}(\mu,\mu,\Omega)$, $W^{1,p}_{loc}(\mu,\mu,\Omega)$ as defined in 2.2.1. A pair $(u,\nabla u) \in W^{1,p}_{loc}(\mu,\Omega)$ is said to be a weak solution of (3.3.1) if

$$\sum_i \int \nabla\Phi_i \cdot A_i(x,u,\nabla u) + \Phi_i B(x,u,\nabla u) = 0$$

for all $\Phi \in W^{1,p}_0(\mu,\Omega)$.

It will be assumed that the following Sobolev inequalities hold for p, q, s with $1 < q < p < s$.

$$\Big(\frac{1}{\omega(B)} \int_B |\Phi - \frac{1}{\omega(B)} \int_B \Phi d\omega|^p d\omega\Big)^{1/p} \leq s_1 r\Big(\frac{1}{\omega(B)} \int_B |\nabla\phi|^q d\omega\Big)^{1/q} \quad (3.3.4)$$

for all $\phi \in W^{1,p}_{loc}(u,\Omega)$ and all balls B with diameter r, $\bar{B} \subseteq \Omega$;

$$\Big(\frac{1}{\omega(B)} \int_B |\Phi - \frac{1}{\omega(B)} \int_B \Phi d\omega|^s d\omega\Big)^{1/s} \leq s_2 r\Big(\frac{1}{\omega(B)} \int_B |\nabla\phi|^p d\omega\Big)^{1/p} \quad (3.3.5)$$

for all $\phi \in W^{1,p}_{loc}(\mu,\Omega)$ and all balls B with diameter r, $\bar{B} \subseteq \Omega$.

It is of course only necessary to assume (3.3.4), (3.3.5) for C^∞ functions and then the usual limit procedures allow general Sobolev functions. Conditions for Sobolev inequalities of the form (3.3.4), (3.3.5) are discussed for scalar-valued functions in Section 2.2.0. The vector-valued case is an obvious corollary. For certain $s > p$, inequality (3.3.5) follows from inequality (3.3.4) as in the proof of Theorem 2.2.56. A simple consequence of inequality (3.3.5) is that if $\Phi \in W^{1,p}_{loc}(\mu,\Omega)$, then $\Phi \in L^s_{loc}(\mu,\Omega)$.

The analysis of the equations will produce a "reverse Hölder" type of maximal function inequality from which the higher integrability of the gradient will follow.

<u>3.3.6 Theorem</u>. If $(u,\nabla u)$ is a weak solution of (3.3.1) in a bounded open set $\Omega \subseteq \mathbf{R}^d$, then there exists $\varepsilon > 0$ so that $|\nabla u| \in L^{p+\varepsilon}_{loc}(\mu,\Omega)$. ε depends only on d, p, q, s, a_0, c_ω, α, s_1, s_2.

<u>Remark</u>. As in [MYE], if $(u,\nabla u) \in W^{1,p}_0(\mu,\Omega)$ and certain weak assumptions are made about $\partial\Omega$, then $|\nabla u| \in L^{p+\varepsilon}(\Omega)$.

<u>Proof</u>. Let Q_∞, Q_1, Q_0 be concentric cubes, $\bar{Q}_0 \subseteq \Omega$ with side lengths S_∞, $2S_\infty$, $3S_\infty$ respectively, as in Section 2.3.0. Estimates will be made over balls $B' = B(x,r)$, $x \in Q_1$, $r < S_\infty/2$, using test functions of the form $\Phi = \phi(u - k)$ where ϕ is a function such that

$$\phi \in C_0^\infty(B(x,2r)),\quad 0 \leq \phi \leq 1,\quad \phi = 1 \text{ on } B',\ |\nabla\phi| \leq 2/r\ . \tag{3.3.7}$$

These calculations will yield the inequality

$$M_R(g^{\bar{q}})(y) \leq cM^{\bar{q}}(g)(y) + \frac{1}{2} M(g^{\bar{q}})(y) + F(y) \tag{3.3.8}$$

for $y \in Q_1$ and $R = S_\infty/2$ where M_R, M are as defined in Section 2.3.0, $g = |\nabla u|^q$, $\bar{q} = p/q$ and c depends only on a_0, s_1, s_2, p, d, c_ω; and $F \in L^{1+\alpha'}(M,Q_1)$ for some $\alpha' > 0$, α' dependent only on α, s, p.

Using propositions 1.1.3, 1.1.4, 1.1.5 and 1.1.9 to show that $F \leq M(F)$ a.e. and letting $f = F^{1/q}$ it follows that

$$M_R(g^{\bar{q}}) \leq cM^{\bar{q}}(g) + M(f^{\bar{q}}) + \frac{1}{2} M(g^{\bar{q}}) \quad \text{a.e. in } Q_1\ .$$

Applying Theorem 2.3.3 it is clear that $|\nabla u| \in L^{p+\varepsilon}(Q_\infty)$ for some $\varepsilon > 0$ dependent only on d, p, q, s, a_0, c_ω, α, s_1, s_2, so $|\nabla u| \in L^{p+\varepsilon}_{loc}(\mu,\Omega)$.

To prove 3.3.8 let $K = \omega^{-1}(B) \int_B u\,d\omega$, $B = B(x,2r)$, $v = u - k$ and $\Phi = \phi^p v$, ϕ as above. Take

$$\nabla\Phi = p\phi^{p-1}\nabla\phi v + \phi^p \nabla v$$

so that $(\Phi,\nabla\Phi) \in W_0^{1,p}(\mu,\Omega)$ by Proposition 2.2.2, and

$$\sum_i \int [p\phi^{p-1} v_i \nabla\phi\cdot A_i + \phi^p \nabla v_i\cdot A_i + \phi^p v_i B_i] = 0\ .$$

Rearranging terms and using the structure conditions (3.3.2), it follows that

$$\int \phi^p |\nabla u|^p \mu \leq \int \phi^p c_1 \mu + a_0 p \int \phi^{p-1}|\nabla\phi||v||\nabla u|^{p-1}\mu$$

$$+ p\int \phi^{p-1}|\nabla\phi||v|a_1\mu + \int \phi^p |v||\nabla u|^{p-1} b_1 \mu + \int \phi^p |v| b_2 \mu\ .$$

Younges' inequality implies that

$$\phi^{p-1}|\nabla\phi||v||\nabla u|^{p-1} \leq \varepsilon\phi^p|\nabla u|^p + \varepsilon^{-(p-1)}|\nabla\phi|^p|v|^p$$

$$|v||\nabla u|^{p-1}b_1 \leq \varepsilon|\nabla u|^p + \varepsilon^{-(p-1)}|v|^p b_1^p\ .$$

Applying these with $\varepsilon = \min\{\frac{1}{4}, (4a_0p)^{-1}\}$, absorbing the gradient terms into the term on the left-hand side, using Hölder's inequality on three of the remaining terms and recalling (3.3.7), it follows that

$$\int_{B'} |\nabla u|^p \mu \le \int_B c_1 \mu + cr^{-p} \int_B |v|^p \mu$$

$$+ cr^{-1}\Big(\int_B |v|^s \mu\Big)^{1/s}\Big(\int_B a_1^{s/(s-1)} \mu\Big)^{(s-1)/s}$$

$$+ c\Big(\int_B |v|^t \mu\Big)^{p/t}\Big(\int_B b_1^{pt/(t-p)} \mu\Big)^{(t-p)/t}$$

$$+ \Big(\int_B |v|^t \mu\Big)^{1/t}\Big(\int_B b_2^{t/(t-1)} \mu\Big)^{(t-1)/t}$$

for some $t < s$ such that $\frac{pt}{t-1} < \frac{ps}{s-1} + \alpha$ and $\frac{t}{t-1} < \frac{s}{s-1} + \alpha$

Finally, use inequality (3.3.4) on the second term, inequality (3.3.5) on the third, followed by an application of Younges' inequality and recall that $\omega(B) \le cC_\omega \omega(B')$ to get that

$$\frac{1}{\omega(B')} \int_{B'} |\nabla u|^p \mu \le \frac{1}{\omega(B)} \int_B c_1 \mu + c\Big(\frac{1}{\omega(B)} \int_B |\nabla u|^q \mu\Big)^{p/q}$$

$$+ \varepsilon \frac{1}{\omega(B)} \int_B |\nabla u|^p \mu$$

$$+ c\Big(\frac{1}{\omega(B)} \int_B a_1^{s/(s-1)} \mu\Big)^{p(s-1)/(p-1)s}$$

$$+ c\Big(\frac{1}{\omega(B)} \int_B |v|^t \mu\Big)^{p/t}\Big(\frac{1}{\omega(B)} \int_B b_1^{pt/(t-p)} \mu\Big)^{(t-p)/t}$$

$$+ \Big(\frac{1}{\omega(B)} \int_B |v|^t \mu\Big)^{1/t}\Big(\frac{1}{\omega(B)} \int_B b_2^{t/(t-1)} \mu\Big)^{(t-1)/t} .$$

Taking the supremum over r, $0 < r \le S_\infty/2$ and letting $\varepsilon = 1/2$, $|\nabla u|^q = g$, $\bar{q} = p/q$ and $R = S_\infty/2$ it is clear that 3.3.8 is verified with

$$F(y) = c(M(c_1)(y) + M^{p(s-1)/(p-1)s}(a_1^{s/(s-1)})(y)$$

$$+ M^{p/t}(|v|^t)(y)M^{(t-p)/t}(b_1^{pt/(t-p)})(y)$$

$$+ M^{1/t}(|v|^t)(y)M^{(t-1)/t}(b_2^{t/(t-1)})(y) ,$$

and $F \in L^{1+\alpha'}(\mu,\Omega_1)$ for some $\alpha' > 0$ because of Propositions 1.1.3, 1.1.4, Assumption 3.3.3, and the fact that $|v| \in L^s_{loc}(\mu,\Omega)$, which follows from inequality 3.3.5. The proof is completed as in 3.3.8. ∎

Continuity of Solutions

Suppose the measure ω is fixed and there is a critical exponent p_0 where the constant $K(x,r)$ in the Sobolev inequality (3.2.12) is bounded for $p > p_0$ and unbounded for $p < p_0$. (If ω is Lebesgue measure, then $p_0 = d$.) If a_0 and α (as in (3.3.2), (3.3.3)) are fixed and only equations satisfying this structure are considered, then under fairly general circumstances the ε given in Theorem 3.3.6 will depend continuously on p so that estimates on the modulus of continuity for solutions u can be derived for $p > p_0 - \varepsilon'$, with $\varepsilon' > 0$ dependent only on p_0, a_0, α, d, and the measure ω. For simplicity only the borderline case $p = p_0$ will be considered.

3.3.9 Corollary. Suppose u, ∇u, Ω, ε are as in Theorem 3.3.6 and

$$\frac{1}{\omega(B)} \int_B |\phi(x) - \frac{1}{\omega(B)} \int_B \phi(y)d\omega(y)|d\omega(x) \tag{3.3.10}$$

$$< K(\int_B |\nabla\phi|^{p_0} d\omega)^{1/p_0}$$

for all $\phi \in W^{1,p}_{loc}(\mu,\mu,\Omega)$ and all balls B with $\bar{B} \subseteq \Omega$. If x, y are Lebesgue points of u with respect to ω, such that $\overline{B(x_0,R)} \subseteq \Omega$ for $x_0 = \frac{x+y}{2}$ and $R = \frac{|x-y|}{2}$, then

$$|u(x)-u(y)| < 8\sqrt{d}\left(\frac{p_0 + 2\varepsilon}{\varepsilon}\right)K\left(\int_{B(x_0,R)} |\nabla u|^{p_0+\varepsilon} d\omega\right)^{1/(p_0+\varepsilon)} \tag{3.3.11}$$

$$\times\ \omega^{\varepsilon/(p_0+\varepsilon)}(B(x_0,R)) .$$

Remark. Conditions for (3.3.10) to hold are discussed in the remark after Theorem 3.2.11.

Example. In the example developed after Theorem 3.2.11, it is easy to see that the critical exponent is d if $\alpha < 0$ and $d + \alpha$ if $\alpha > 0$. Also, since $\omega(B(x,r)) \sim r^d \max^\alpha\{r,\mathrm{dist}(x,K)\}$, $\alpha > -\gamma > d$, it follows from Corollary 3.3.9 that Hölder continuity can be established for solutions in the borderline cases.

<u>Proof of Corollary 3.3.9.</u> $u = (u_1,\ldots,u_d)$ so

$$\frac{1}{\omega(B)} \int_B |u_i(x) - \frac{1}{\omega(B)} \int_B u_i(y)d\omega(y)|d\omega(x)$$

$$\leq K(\int_B |\nabla u_i|^{p_0} d\omega)^{1/p_0}$$

$$\leq K\omega^{\varepsilon/(p_0+\varepsilon)}(B)(\int_B |\nabla u_i|^{p_0+\varepsilon} d\omega)^{1/(p_0+\varepsilon)} .$$

Given $x,y \in \Omega$, Lebesgue points of u with respect to ω, and a ball $B(x_0,R)$ such that $\overline{B(x_0,R)} \subseteq \Omega$ and $x,y \in B(x_0,R)$, use the geometry described in the first remark after Theorem 3.2.5 and apply the second part of Theorem 3.2.5 with

$$f(x) = K(\int_{B(x_0,R)} |\nabla u|^{p_0+\varepsilon} d\omega)^{1/(p_0+\varepsilon)} x^{\varepsilon/p_0+\varepsilon}$$

to conclude that

$$|u_i(x) - u_i(y)|$$

$$\leq 8(\frac{p + 2\varepsilon}{\varepsilon})K(\int_B |\nabla u|^{p_0+\varepsilon} d\omega)^{1/(p_0+\varepsilon)} \omega^{\varepsilon/(p_0+\varepsilon)}(B(x_0,R)) .$$

Now let $x_0 \to \frac{x+y}{2}$ and $R \downarrow \frac{|x-y|}{2}$ in such a way that x, y remain in $B(x_0,R)$ so that (3.3.11) is verified. ∎

REFERENCES

[A1] D. R. Adams, Traces and potentials arising from translation invariant operators, Ann. Scuola Norm. Sup., Pisa 25 (1971), 203-207.

[A2] D. R. Adams, A trace inequality for generalized potentials Studia Math. 48 (1973), 99-105.

[A3] D. R. Adams, On the existence of capacitary strong type estimates in R^n, Ark. Mat. 14 (1976), 125-140.

[AS] D. G. Aronson and James Serrin, Local behavior of solutions of quasilinear parabolic equations, Arch. Rat. Mech. Anal. 25 (1967), 81-122.

[AR] M. Artola, unpublished manuscript.

[BG] T. Bagby, Quasi topologies and rational approximation, J. Functional Analysis 10, No. 3 (1972), 259-268.

[B] A. S. Besicovitch, A general form of the covering principle and relative differentiation of additive functions, Proc. Cambridge Philos. Soc. 42 (1946), 1-10.

[BI] E. Bombieri, Theory of minimal surfaces and a counterexample to the Bernstein conjecture in high dimensions, Mimeographed Notes of Lectures held at Courant Institute, New York University, 1970.

[BR] J. S. Bradley, Hardy inequalities with mixed norms, Canad. Math. Bull. 21, No. 4 (1978), 405-408.

[CA] S. Campanato, Proprieta di una famiglia di spazi funzionali, Ann. Scuola Norm. Sup. Pisa 18, No. 3 (1964), 137-160.

[C] G. Choquet, Theory of capacities, Ann. Inst. Fourier (Grenoble) 5 (1953-1954), 131-295.

[EP] D. E. Edmunds and L. A. Peletier, A. Harnack inequality for weak solutions of degenerate quasilinear elliptic equations, J. London Math. Soc. 5, No. 2 (1972), 21-31.

[FJK] E. B. Fabes, D. S. Jerison, C. E. Kenig, The Wiener test for degenerate elliptic equations, to appear.

[FKS] E. B. Fabes, C. E. Kenig, R. P. Serapioni, The local regularity of solutions of degenerate elliptic equations, Comm. P.D.E.s 7 (1) 1982, 77-116.

[F1] H. Federer, The area of a nonparametric surface, Proc. Amer. Math. Soc. 2, No. 3 (1960), 436-439.

[F2] H. Federer, Geometric measure theory, Springer-Verlag, 1969.

[GZ] R. Gariepy and W. P. Ziemer, Behavior at the boundary of solutions of quasilinear elliptic equations, Arch. Rat. Mech. Anal. 67 (1977), 25-39.

[GH] F. W. Gehring, The L^p-integrability of the partial derivatives of a quasiconformal mapping, Acta Math. 130 (1973), 265-277.

[GI] M. Giaquinta, Multiple integrals in the calculus of variations and nonlinear elliptic systems, Princeton Univ. Press, 1983.

[GM] M. Giaquinta, G. Modica, Regularity results for some classes of higer order nonlinear elliptic systems, J. für Reine U. Angew. Math. 311-312 (1979), 145-169.

[GU] W. Gustin, The Boxing inequality, J. Math. Mech. 9 (1960), 229-239.

[G] M. de Guzmán, Differentiation of integrals in R^n, Lecture Notes in Mathematics 481, Springer-Verlag.

[HS] E. Hewitt and K. Stromberg, Real and abstract analysis, Springer-Verlag, 1965.

[JN] F. John and L. Nirenberg, On fucntions of bounded mean oscillation, Comm. Pure Appl. Math. 14 (1961), 415-426.

[K] S. N. Kruzkov, Certain properties of solutions to elliptic equations, Soviet Math. 4 (1963), 686-695.

[MA1] V. G. Maz'ya, A theorem on the multidimensional Schröedinger operator, (Russian), Izv. Akad. Nak. 28 (1964), 1145-1172.

[MA2] V. G. Maz'ya, On some integral inequalities for functions of several variables, (Russian), Problems in Math. Analysis, No. 3 (1972), Leningrad U.

[MYE] N. G. Meyers and A. Elcrat, Some results on regularity for solutions of nonlinear elliptic systems and quasi-regular functions, Duke Math. J. 42 (1975), 121-136.

[MY1] N. G. Meyers, Integral inequalities of Poincaré and Wirtinger type, Arch. Rat. Mech. Anal. 68 (1978), 113-120.

[MY2] N. G. Meyers, Mean oscillation over cubes and Hölder continuity, Proc. Amer. Math. Soc. 15 (1964), 717-721.

[MR] A. P. Morse, Perfect blankets, Trans. Amer. Math. Soc. 6 (1947) 418-442.

[ME1] J. Moser, On Harnack's theorem for elliptic differential equations, Comm. Pure Appl. Math. 14 (1961), 577-591.

[ME2] J. Moser, On a pointwise estimate for parabolic differential equations, Comm. Pure Appl. Math. 24 (1971), 727-740.

[M1] B. Muckenhoupt, The equivalence of two conditions for weight functions, Studia Math. 49 (1974), 101-106.

[M2] B. Muckenhoupt, Hardy's inequality with weights, Studia Math. 34 (1972), 31-38.

[MW] B. Muckenhoupt and R. L. Wheedan, Weighted norm inequalities for fractional integrals, Trans. Amer. Math. Soc. 192 (1974), 261-274.

[MS] M. K. V. Murthy and G. Stampacchia, Boundary value problems for some degenerate elliptic operators, Annali di Matematica 80 (1968), 1-122.

[SM] P. D. Smith, A regularity theorem for a singular elliptic equation, Applicable Analysis, 14 #3 (1983), 223-236.

[ST] E. M. Stein, Singular integrals and differentiality properties of functions, Princeton U. Press, 1970.

[S1] E. W. Stredulinsky, Higher integrability from reverse Hölder inequalities, Indiana Univ. Math. J. 29 (1980), 407-413.

[S2] E. W. Stredulinsky, Weighted inequalities and applications to degenerate partial differential equations. Doctoral thesis 1981, Indiana University.

[TL] G. Talenti, Osservazioni supra una classe di disuguaglinze, Rend. Sem. Mat. e Fis. Milano 39 (1969), 171-185.

[TM] G. Tomaselli, A class of inequalities, Boll. Un. Mat. Ital. 21 (1969), 622-631.

[T1] N. S. Trudinger, On the regularity of generalized solutions of linear, nonuniformly elliptic equations, Arch. Rat. Mech. Anal. 41 (1971), 50-62.

[T2] N. S. Trudinger, Linear elliptic operators with measurable coefficients, Ann. Scuola Norm. Sup. Pisa 27 (1973), 265-308.

[T3] N. S. Trudinger, On Harnack inequalities and their application to quasilinear elliptic equations, Comm. Pure Appl. Math. 20 (1967), 721-747.

[W] G. V. Welland, Weighted norm inequalities for fractional integrals, Proc. Amer. Math. Soc. 51 (1975), 143-148.

[W1] K.-O. Widman, Hölder continuity of solutions of elliptic systems, Manuscripta Math. 5 (1971), 299-308.

INDEX

Vol. 926: Geometric Techniques in Gauge Theories. Proceedings, 1981. Edited by R. Martini and E.M.de Jager. IX, 219 pages. 1982.

Vol. 927: Y. Z. Flicker, The Trace Formula and Base Change for GL (3). XII, 204 pages. 1982.

Vol. 928: Probability Measures on Groups. Proceedings 1981. Edited by H. Heyer. X, 477 pages. 1982.

Vol. 929: Ecole d'Eté de Probabilités de Saint-Flour X – 1980. Proceedings, 1980. Edited by P.L. Hennequin. X, 313 pages. 1982.

Vol. 930: P. Berthelot, L. Breen, et W. Messing, Théorie de Dieudonné Cristalline II. XI, 261 pages. 1982.

Vol. 931: D.M. Arnold, Finite Rank Torsion Free Abelian Groups and Rings. VII, 191 pages. 1982.

Vol. 932: Analytic Theory of Continued Fractions. Proceedings, 1981. Edited by W.B. Jones, W.J. Thron, and H. Waadeland. VI, 240 pages. 1982.

Vol. 933: Lie Algebras and Related Topics. Proceedings, 1981. Edited by D. Winter. VI, 236 pages. 1982.

Vol. 934: M. Sakai, Quadrature Domains. IV, 133 pages. 1982.

Vol. 935: R. Sot, Simple Morphisms in Algebraic Geometry. IV, 146 pages. 1982.

Vol. 936: S.M. Khaleelulla, Counterexamples in Topological Vector Spaces. XXI, 179 pages. 1982.

Vol. 937: E. Combet, Intégrales Exponentielles. VIII, 114 pages. 1982.

Vol. 938: Number Theory. Proceedings, 1981. Edited by K. Alladi. IX, 177 pages. 1982.

Vol. 939: Martingale Theory in Harmonic Analysis and Banach Spaces. Proceedings, 1981. Edited by J.-A. Chao and W.A. Woyczyński. VIII, 225 pages. 1982.

Vol. 940: S. Shelah, Proper Forcing. XXIX, 496 pages. 1982.

Vol. 941: A. Legrand, Homotopie des Espaces de Sections. VII, 132 pages. 1982.

Vol. 942: Theory and Applications of Singular Perturbations. Proceedings, 1981. Edited by W. Eckhaus and E.M. de Jager. V, 363 pages. 1982.

Vol. 943: V. Ancona, G. Tomassini, Modifications Analytiques. IV, 120 pages. 1982.

Vol. 944: Representations of Algebras. Workshop Proceedings, 1980. Edited by M. Auslander and E. Lluis. V, 258 pages. 1982.

Vol. 945: Measure Theory. Oberwolfach 1981, Proceedings. Edited by D. Kölzow and D. Maharam-Stone. XV, 431 pages. 1982.

Vol. 946: N. Spaltenstein, Classes Unipotentes et Sous-groupes de Borel. IX, 259 pages. 1982.

Vol. 947: Algebraic Threefolds. Proceedings, 1981. Edited by A. Conte. VII, 315 pages. 1982.

Vol. 948: Functional Analysis. Proceedings, 1981. Edited by D. Butković, H. Kraljević, and S. Kurepa. X, 239 pages. 1982.

Vol. 949: Harmonic Maps. Proceedings, 1980. Edited by R.J. Knill, M. Kalka and H.C.J. Sealey. V, 158 pages. 1982.

Vol. 950: Complex Analysis. Proceedings, 1980. Edited by J. Eells. IV, 428 pages. 1982.

Vol. 951: Advances in Non-Commutative Ring Theory. Proceedings, 1981. Edited by P.J. Fleury. V, 142 pages. 1982.

Vol. 952: Combinatorial Mathematics IX. Proceedings, 1981. Edited by E. Billington, S. Oates-Williams, and A.P. Street. XI, 443 pages. 1982.

Vol. 953: Iterative Solution of Nonlinear Systems of Equations. Proceedings, 1982. Edited by R. Ansorge, Th. Meis, and W. Törnig. VII, 202 pages. 1982.

Vol. 954: S.G. Pandit, S.G. Deo, Differential Systems Involving Impulses. VII, 102 pages. 1982.

Vol. 955: G. Gierz, Bundles of Topological Vector Spaces and Their Duality. IV, 296 pages. 1982.

Vol. 956: Group Actions and Vector Fields. Proceedings, 1981. Edited by J.B. Carrell. V, 144 pages. 1982.

Vol. 957: Differential Equations. Proceedings, 1981. Edited by D.G. de Figueiredo. VIII, 301 pages. 1982.

Vol. 958: F.R. Beyl, J. Tappe, Group Extensions, Representations, and the Schur Multiplicator. IV, 278 pages. 1982.

Vol. 959: Géométrie Algébrique Réelle et Formes Quadratiques, Proceedings, 1981. Edité par J.-L. Colliot-Thélène, M. Coste, L. Mahé, et M.-F. Roy. X, 458 pages. 1982.

Vol. 960: Multigrid Methods. Proceedings, 1981. Edited by W. Hackbusch and U. Trottenberg. VII, 652 pages. 1982.

Vol. 961: Algebraic Geometry. Proceedings, 1981. Edited by J.M. Aroca, R. Buchweitz, M. Giusti, and M. Merle. X, 500 pages. 1982.

Vol. 962: Category Theory. Proceedings, 1981. Edited by K.H. Kamps, D. Pumplün, and W. Tholen, XV, 322 pages. 1982.

Vol. 963: R. Nottrot, Optimal Processes on Manifolds. VI, 124 pages. 1982.

Vol. 964: Ordinary and Partial Differential Equations. Proceedings, 1982. Edited by W.N. Everitt and B.D. Sleeman. XVIII, 726 pages. 1982.

Vol. 965: Topics in Numerical Analysis. Proceedings, 1981. Edited by P.R. Turner. IX, 202 pages. 1982.

Vol. 966: Algebraic K-Theory. Proceedings, 1980, Part I. Edited by R.K. Dennis. VIII, 407 pages. 1982.

Vol. 967: Algebraic K-Theory. Proceedings, 1980. Part II. VIII, 409 pages. 1982.

Vol. 968: Numerical Integration of Differential Equations and Large Linear Systems. Proceedings, 1980. Edited by J. Hinze. VI, 412 pages. 1982.

Vol. 969: Combinatorial Theory. Proceedings, 1982. Edited by D. Jungnickel and K. Vedder. V, 326 pages. 1982.

Vol. 970: Twistor Geometry and Non-Linear Systems. Proceedings, 1980. Edited by H.-D. Doebner and T.D. Palev. V, 216 pages. 1982.

Vol. 971: Kleinian Groups and Related Topics. Proceedings, 1981. Edited by D.M. Gallo and R.M. Porter. V, 117 pages. 1983.

Vol. 972: Nonlinear Filtering and Stochastic Control. Proceedings, 1981. Edited by S.K. Mitter and A. Moro. VIII, 297 pages. 1983.

Vol. 973: Matrix Pencils. Proceedings, 1982. Edited by B. Kågström and A. Ruhe. XI, 293 pages. 1983.

Vol. 974: A. Draux, Polynômes Orthogonaux Formels – Applications. VI, 625 pages. 1983.

Vol. 975: Radical Banach Algebras and Automatic Continuity. Proceedings, 1981. Edited by J.M. Bachar, W.G. Bade, P.C. Curtis Jr., H.G. Dales and M.P. Thomas. VIII, 470 pages. 1983.

Vol. 976: X. Fernique, P.W. Millar, D.W. Stroock, M. Weber, Ecole d'Eté de Probabilités de Saint-Flour XI – 1981. Edited by P.L. Hennequin. XI, 465 pages. 1983.

Vol. 977: T. Parthasarathy, On Global Univalence Theorems. VIII, 106 pages. 1983.

Vol. 978: J. Ławrynowicz, J. Krzyż, Quasiconformal Mappings in the Plane. VI, 177 pages. 1983.

Vol. 979: Mathematical Theories of Optimization. Proceedings, 1981. Edited by J.P. Cecconi and T. Zolezzi. V, 268 pages. 1983.

Vol. 980: L. Breen. Fonctions thêta et théorème du cube. XIII, 115 pages. 1983.

Vol. 981: Value Distribution Theory. Proceedings, 1981. Edited by I. Laine and S. Rickman. VIII, 245 pages. 1983.

Vol. 982: Stability Problems for Stochastic Models. Proceedings, 1982. Edited by V. V. Kalashnikov and V. M. Zolotarev. XVII, 295 pages. 1983.

Vol. 983: Nonstandard Analysis-Recent Developments. Edited by A. E. Hurd. V, 213 pages. 1983.

Vol. 984: A. Bove, J. E. Lewis, C. Parenti, Propagation of Singularities for Fuchsian Operators. IV, 161 pages. 1983.

Vol. 985: Asymptotic Analysis II. Edited by F. Verhulst. VI, 497 pages. 1983.

Vol. 986: Séminaire de Probabilités XVII 1981/82. Proceedings. Edited by J. Azéma and M. Yor. V, 512 pages. 1983.

Vol. 987: C. J. Bushnell, A. Fröhlich, Gauss Sums and p-adic Division Algebras. XI, 187 pages. 1983.

Vol. 988: J. Schwermer, Kohomologie arithmetisch definierter Gruppen und Eisensteinreihen. III, 170 pages. 1983.

Vol. 989: A. B. Mingarelli, Volterra-Stieltjes Integral Equations and Generalized Ordinary Differential Expressions. XIV, 318 pages. 1983.

Vol. 990: Probability in Banach Spaces IV. Proceedings, 1982. Edited by A. Beck and K. Jacobs. V, 234 pages. 1983.

Vol. 991: Banach Space Theory and its Applications. Proceedings, 1981. Edited by A. Pietsch, N. Popa and I. Singer. X, 302 pages. 1983.

Vol. 992: Harmonic Analysis, Proceedings, 1982. Edited by G. Mauceri, F. Ricci and G. Weiss. X, 449 pages. 1983.

Vol. 993: R. D. Bourgin, Geometric Aspects of Convex Sets with the Radon-Nikodým Property. XII, 474 pages. 1983.

Vol. 994: J.-L. Journé, Calderón-Zygmund Operators, Pseudo-Differential Operators and the Cauchy Integral of Calderón. VI, 129 pages. 1983.

Vol. 995: Banach Spaces, Harmonic Analysis, and Probability Theory. Proceedings, 1980–1981. Edited by R.C. Blei and S.J. Sidney. V, 173 pages. 1983.

Vol. 996: Invariant Theory. Proceedings, 1982. Edited by F. Gherardelli. V, 159 pages. 1983.

Vol. 997: Algebraic Geometry – Open Problems. Edited by C. Ciliberto, F. Ghione and F. Orecchia. VIII, 411 pages. 1983.

Vol. 998: Recent Developments in the Algebraic, Analytical, and Topological Theory of Semigroups. Proceedings, 1981. Edited by K.H. Hofmann, H. Jürgensen and H. J. Weinert. VI, 486 pages. 1983.

Vol. 999: C. Preston, Iterates of Maps on an Interval. VII, 205 pages. 1983.

Vol. 1000: H. Hopf, Differential Geometry in the Large, VII, 184 pages. 1983.

Vol. 1001: D.A. Hejhal, The Selberg Trace Formula for PSL(2, ℝ). Volume 2. VIII, 806 pages. 1983.

Vol. 1002: A. Edrei, E.B. Saff, R.S. Varga, Zeros of Sections of Power Series. VIII, 115 pages. 1983.

Vol. 1003: J. Schmets, Spaces of Vector-Valued Continuous Functions. VI, 117 pages. 1983.

Vol. 1004: Universal Algebra and Lattice Theory. Proceedings, 1982. Edited by R. S. Freese and O. C. Garcia. VI, 308 pages. 1983.

Vol. 1005: Numerical Methods. Proceedings, 1982. Edited by V. Pereyra and A. Reinoza. V, 296 pages. 1983.

Vol. 1006: Abelian Group Theory. Proceedings, 1982/83. Edited by R. Göbel, L. Lady and A. Mader. XVI, 771 pages. 1983.

Vol. 1007: Geometric Dynamics. Proceedings, 1981. Edited by J. Palis Jr. IX, 827 pages. 1983.

Vol. 1008: Algebraic Geometry. Proceedings, 1981. Edited by J. Dolgachev. V, 138 pages. 1983.

Vol. 1009: T. A. Chapman, Controlled Simple Homotopy Theory and Applications. III, 94 pages. 1983.

Vol. 1010: J.-E. Dies, Chaînes de Markov sur les permutations. IX, 226 pages. 1983.

Vol. 1011: J. M. Sigal. Scattering Theory for Many-Body Quantum Mechanical Systems. IV, 132 pages. 1983.

Vol. 1012: S. Kantorovitz, Spectral Theory of Banach Space Operators. V, 179 pages. 1983.

Vol. 1013: Complex Analysis – Fifth Romanian-Finnish Seminar. Part 1. Proceedings, 1981. Edited by C. Andreian Cazacu, N. Boboc, M. Jurchescu and I. Suciu. XX, 393 pages. 1983.

Vol. 1014: Complex Analysis – Fifth Romanian-Finnish Seminar. Part 2. Proceedings, 1981. Edited by C. Andreian Cazacu, N. Boboc, M. Jurchescu and I. Suciu. XX, 334 pages. 1983.

Vol. 1015: Equations différentielles et systèmes de Pfaff dans le champ complexe – II. Seminar. Edited by R. Gérard et J. P. Ramis. V, 411 pages. 1983.

Vol. 1016: Algebraic Geometry. Proceedings, 1982. Edited by M. Raynaud and T. Shioda. VIII, 528 pages. 1983.

Vol. 1017: Equadiff 82. Proceedings, 1982. Edited by H. W. Knobloch and K. Schmitt. XXIII, 666 pages. 1983.

Vol. 1018: Graph Theory, Łagów 1981. Proceedings, 1981. Edited by M. Borowiecki, J. W. Kennedy and M. M. Sysło. X, 289 pages. 1983.

Vol. 1019: Cabal Seminar 79–81. Proceedings, 1979–81. Edited by A. S. Kechris, D. A. Martin and Y. N. Moschovakis. V, 284 pages. 1983.

Vol. 1020: Non Commutative Harmonic Analysis and Lie Groups. Proceedings, 1982. Edited by J. Carmona and M. Vergne. V, 187 pages. 1983.

Vol. 1021: Probability Theory and Mathematical Statistics. Proceedings, 1982. Edited by K. Itô and J.V. Prokhorov. VIII, 747 pages. 1983.

Vol. 1022: G. Gentili, S. Salamon and J.-P. Vigué. Geometry Seminar "Luigi Bianchi", 1982. Edited by E. Vesentini. VI, 177 pages. 1983.

Vol. 1023: S. McAdam, Asymptotic Prime Divisors. IX, 118 pages. 1983.

Vol. 1024: Lie Group Representations I. Proceedings, 1982–1983. Edited by R. Herb, R. Lipsman and J. Rosenberg. IX, 369 pages. 1983.

Vol. 1025: D. Tanré, Homotopie Rationnelle: Modèles de Chen, Quillen, Sullivan. X, 211 pages. 1983.

Vol. 1026: W. Plesken, Group Rings of Finite Groups Over p-adic Integers. V, 151 pages. 1983.

Vol. 1027: M. Hasumi, Hardy Classes on Infinitely Connected Riemann Surfaces. XII, 280 pages. 1983.

Vol. 1028: Séminaire d'Analyse P. Lelong – P. Dolbeault – H. Skoda. Années 1981/1983. Edité par P. Lelong, P. Dolbeault et H. Skoda. VIII, 328 pages. 1983.

Vol. 1029: Séminaire d'Algèbre Paul Dubreil et Marie-Paule Malliavin. Proceedings, 1982. Edité par M.-P. Malliavin. V, 339 pages. 1983.

Vol. 1030: U. Christian, Selberg's Zeta-, L-, and Eisensteinseries. XII, 196 pages. 1983.

Vol. 1031: Dynamics and Processes. Proceedings, 1981. Edited by Ph. Blanchard and L. Streit. IX, 213 pages. 1983.

Vol. 1032: Ordinary Differential Equations and Operators. Proceedings, 1982. Edited by W. N. Everitt and R. T. Lewis. XV, 521 pages. 1983.

Vol. 1033: Measure Theory and its Applications. Proceedings, 1982. Edited by J. M. Belley, J. Dubois and P. Morales. XV, 317 pages. 1983.

Vol. 1034: J. Musielak, Orlicz Spaces and Modular Spaces. V, 222 pages. 1983.